RONALD ORENSTEIN

HOW ON EARTH?

A QUESTION-AND-ANSWER BOOK ABOUT HOW OUR PLANET WORKS

KPk
Key Porter Kids

For Randy and Jenny

Photography Credits

© Andre Bartschi/Planet Earth Pictures, 60 left; © Brian Beck, 5, 44-45; © Chris Howes/Planet Earth Pictures, 54; © J.A. Kraulis, 33 bottom, 34, 62; © Ford Kristo/Planet Earth Pictures, 20 left; © John Lythgoe/Planet Earth Pictures, 23 bottom; NASA (Courtesy Roberta Bondar), 10, 11 top, 11 bottom; © M. Timothy O'Keefe/Bruce Coleman Inc., 59; © Ronald Orenstein, 38; © Doug Perrine/Planet Earth Pictures, 43 bottom; © Keith Scholey/Planet Earth Pictures, 60 right, 63 bottom; © Peter Scoones/Planet Earth Pictures, 78; © P.P. Shirshov Institute, 46; © Flip Schulke/Planet Earth Pictures, 43 top; © Tony Stone Images, cover, 2; © Kim Taylor/Bruce Coleman Inc., 61 left, 61 right; © Ron and Valerie Taylor/ Bruce Coleman Inc., 79 right.

The remaining copyrighted photographs were supplied by First Light Associated Photographers: Kelvin Aitken, 47 bottom; Grant Black, 86; John Cancalosi, 68 right; D. Cavagnaro, 40 right, 76; Vic Cox, 30; David J. Cross, 28; James Dennis, 50-51; George D. Dodge, Kathleen M. Dodge, Dale R. Thompson, 4, 47 top, 55 top, 57 bottom, 63 top, 64, 73, 77 bottom; Bud Freund, 56; D. and J. Heaton, 5, 8-9, 24; Stephen Homer, 18 inset (top), 18 inset (bottom), 69 top, 87 right, 92; Keith Kent, 7, 41 bottom, 70, 79 center; Thomas Kitchin, 80; Jerry Koba-lenko, 41 top; Dennis Kunkel, 48; Y. Levy, 52-53; Franz Maier, 1, 81; Luiz Marigo, 90 left, 91; Mary Ellen McQuay, 65, 65 inset; Brian Milne, 40 left, 55 bottom, 71; Gene Moore, 42; Warren Morgan, 32; Pat Morrow, 23 top, 35 inset; Guy Motil, 18-19; David Nunuk, 87 left; Charles O'Rear, 29 right; Phototake 2, 16; Dave Reede, 5, 84-85; Ed Reschke, 77 top; Jim Richardson, 68 left; Benjamin Rondel, 89; Ken Sakamoto, 26-27, cover; Kevin Schafer, 29 left, 69 bottom; Schafer and Hill, 81 left; Robert Semeniuk, 6, 35; Steve Short, 14-15, 25; Clyde H. Smith, 79 left; Mark Stephenson, 20 right; Paul von Baich, 17; Ron Watts, 33 top, 36-37, 39, 82; Wayne Wegner, 57 top; Darwin R. Wiggett, 22; Jim Zuckerman, 58.

CANADIAN CATALOGUING IN PUBLICATION DATA

Orenstein, Ronald I. (Ronald Isaac), 1946–
    How on earth? : a question-and-answer book about how our planet works

Includes bibliographical references and index.

ISBN 1-55013-513-9

1. Earth — Miscellanea — Juvenile literature. I. Title.

QB631.4.074 1995   j550   C94-932858-8

The publisher gratefully acknowledges the assistance of the Canada Council, the Ontario Arts Council and the Ontario Publishing Centre.

Key Porter Books Limited
70 The Esplanade
Toronto, Ontario
Canada M5E 1R2

Printed and bound in Hong Kong

95 96 97 98 99 6 5 4 3 2 1

Design: Scott Richardson
Typesetting: MacTrix DTP
Illustrations: Angela Vaculik; pages 67, 72 and 73, Brian Franczak

HOW ON EARTH
?

T126244

Contents

OUR PLANET

LIFE ON EARTH

HELPING THE EARTH

Acknowledgments

No one is an expert on everything. While I was writing this book, I was lucky enough to get help from experts on the earth and its creatures. I would like to thank them. They are: Sankar Chatterjee, Desmond Collins, Brian Franczak, Valerius Geist, William K. Hartmann, Christine Janis, Bob Johns of the National Severe Storms Forecast Center, Robert Johnson, Don Lessem, Barry Kent MacKay, Ed Rappaport of the National Hurricane Center, Chat Reynders of the Whale Conservation Institute, Katherine Scott, Paul Sereno, George Shoemaker of Meteor Crater Enterprises, Janet Sumner of Pollution Probe, Janet Waddington, Warren Wagner, and Chris Wiggins. Parts of this book were read over by Faith Campbell, Robert Johnson, D. Morris, Jeheskel Shoshani, and Richard Winterbottom. I also received useful comments from subscribers to the dinosaur, marine mammal and primate mailing lists on the Internet. If I've made any mistakes, though, they are my fault (and that includes leaving anyone out of this list!).

Thanks, too, to Angela Vaculik and, especially, my editor Laurie Coulter, and to the staff at First Light who helped me select the photographs.

My friends and family helped, too. A special thank you to my dear friend Kaaren Dickson and her daughter Kathleen Dickson Overs, who loves frogs. As for my parents Charles and Mary Orenstein, my sister Eve Dexter, and especially my children Randy and Jenny — no thanks can ever be enough. To finish this book, I needed them all.

Introduction

This is a book of questions and answers about the earth and its animals and plants. Some of the questions may be ones that you have asked yourself. Others may be about things you have never heard of before.

Everybody studies nature in their own way. Some people are interested in a favorite animal or plant. Others study the way the earth itself works. Still others study the way life has grown and changed throughout our planet's history.

No matter where you start, your journey through nature is bound to lead you in all sorts of interesting directions, including some that may surprise you.

This book is designed to help you do just that. If you like, you can start at the beginning and read it all the way through. Or you can start anywhere you like, with something that you find interesting. Wherever you start, though, look for questions that are answered on other pages of the book. You can follow these questions back and forth through the book, exploring as you go. You can get to almost every page in the book that way.

But knowing about nature isn't enough anymore. Today, the earth is in danger. As you read this book, keep your eyes open for hints about things you can do to help our planet.

And don't stop there. No one book can tell you everything there is to know about nature. Keep reading and learning. You may be interested in *How on Earth? A Question-and-Answer Book about How Animals and Plants Live*. It is about the lives of animals and plants, and the ways they survive in many different places around the world.

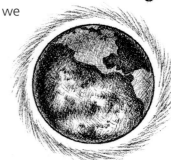

One more thing. There are many questions about nature that we haven't been able to answer yet. You will find a few of them in this book. I hope that some of you will be the ones to answer them!

OUR PLANET

The Earth Is Born

Our earth is part of the solar system. This collection of planets, asteroids, meteors, comets and dust whirls around a medium-sized yellow star called the sun.

The earth is not the biggest of the planets. It is not the smallest, either. In fact, it wouldn't be all that unusual except for three things: it has a very large moon, it has oceans, and — luckily for us — it has life. But 5 000 000 000 years ago it had none of those things. It didn't even exist.

THE MOON AND EARTH

■ WHAT WAS THERE BEFORE THE EARTH FORMED?
Our solar system started out as a cloud of gas and dust. A huge, glowing ball of gas — the sun — formed in the center.
Over the centuries, bits of dust in the cloud collided with each other, forming larger pieces. It was a little like what happens when you make a snowball. As you pack snow onto a snowball, it gets bigger. The snow at the center of the ball turns into a lump of ice.

The bits of dust that stuck together in space turned into rocks.

PAGES 8 AND 9: AURLAND FJORD, NORWAY

The bigger the rocks became, the more gravity they had to pull in dust and other rocks. They grew at last into huge space rocks called planetesimals. It may have taken 100 000 years for the dust cloud to turn into millions of planetesimals the size of mountains.

▲ HOW WAS THE EARTH BORN?
The earth began as a huge rock, or planetesimal. It grew bigger and bigger as more and more rocks smashed into it. It took about 70 million years to reach the size it is today.
As the planetesimals smashed into each other, two things happened —

either they broke each other into pieces, or they stuck together to make a larger planetesimal. The smashed pieces became small rocks called meteors. The larger planetesimals soon got so big that the pull of their gravity swept up smaller rocks in their path, the way a snowball rolling downhill collects more snow. In a few million years, instead of planetesimals, there were millions of meteors and the nine planets: Mercury, Venus, Earth, Mars, Jupiter, Saturn, Uranus, Neptune and Pluto.

WHAT ABOUT...?
Why are there tides?
See page 18.

A VIEW OF JUPITER FROM ITS EQUATOR TO THE SOUTHERN POLAR REGION

■ WHAT DID THE EARLY EARTH LOOK LIKE?

When the earth first formed, it didn't look anything like it does today. Instead of oceans of water, it may have been covered by a sea of molten lava.

As the earth got bigger, the pull of its gravity became stronger. More and more planetesimals smashed into it. They fell so fast that when they hit the earth they exploded. These explosions gave off enough heat to melt the surface of the planet. Scientists think that in those early days the earth may have been covered with a sea of molten lava, perhaps hundreds of kilometers (miles) deep. Rocks brought back from the moon show that it, too, was once covered with lava.

■ WHERE DID THE MOON COME FROM?

The moon may have formed when a giant planetesimal hit the earth, blasting part of the earth's surface into space.

We don't know exactly how the moon was made. Here is the latest theory.

About 50 000 000 years after our planet was born, a giant planetesimal, perhaps the size of the planet Mars, hit the earth. The collision crushed millions of tonnes (tons) of rock into dust and blasted it into space. The dust collected in a ring around the earth. This new dust cloud soon became a swarm of tiny moonlets and the moonlets fused together to form our moon.

NEPTUNE

The Earth's Growth Chart

People once believed that humans had been here almost since the beginning of the world. Today we know that the earth is at least 4 600 000 000 years old and that humans were one of the last creatures to appear on the planet.

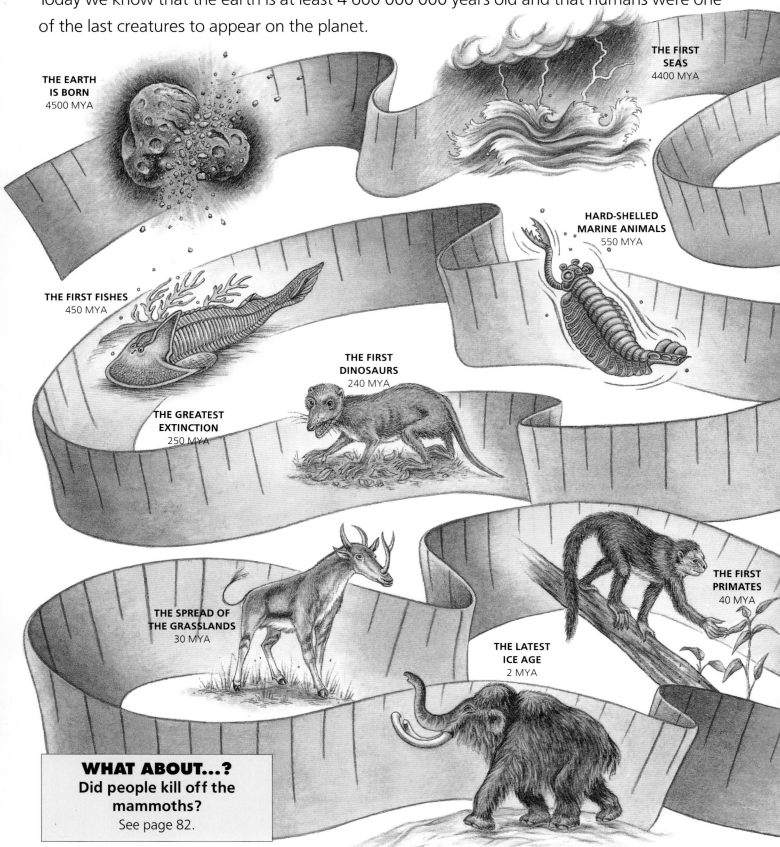

THE EARTH IS BORN
4500 MYA

THE FIRST SEAS
4400 MYA

HARD-SHELLED MARINE ANIMALS
550 MYA

THE FIRST FISHES
450 MYA

THE FIRST DINOSAURS
240 MYA

THE GREATEST EXTINCTION
250 MYA

THE FIRST PRIMATES
40 MYA

THE SPREAD OF THE GRASSLANDS
30 MYA

THE LATEST ICE AGE
2 MYA

WHAT ABOUT...?
Did people kill off the mammoths?
See page 82.

HOW CAN WE TELL HOW OLD THE EARTH IS?

Radioactive atoms provide a "clock" that we can use to measure the age of rocks and other objects.

Some atoms, called isotopes, are radioactive. As they decay, they give off bits of themselves. By measuring how much of its radioactive material has decayed, scientists can tell how old a rock is.

Some isotopes, like uranium-238, take millions of years to decay. They have been used to date meteorite fragments as old as the earth itself. It was not until scientists discovered isotope "clocks" that we learned how old the earth really is.

WHAT ABOUT...?
How are glaciers formed?
See page 34.

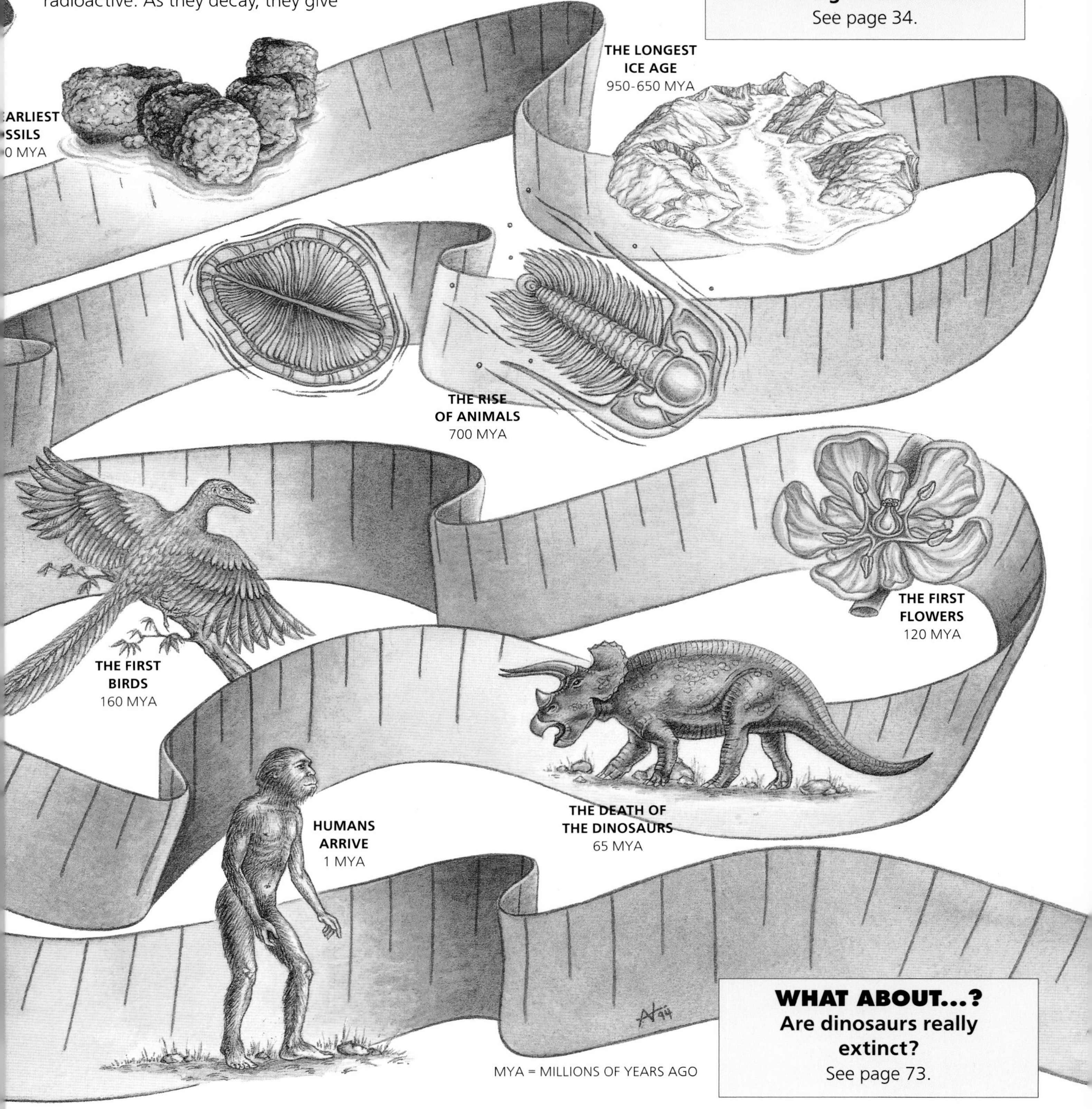

EARLIEST
FOSSILS
0 MYA

THE LONGEST
ICE AGE
950-650 MYA

THE RISE
OF ANIMALS
700 MYA

THE FIRST
FLOWERS
120 MYA

THE FIRST
BIRDS
160 MYA

THE DEATH OF
THE DINOSAURS
65 MYA

HUMANS
ARRIVE
1 MYA

MYA = MILLIONS OF YEARS AGO

WHAT ABOUT...?
Are dinosaurs really extinct?
See page 73.

Sea and Sky

Without oceans, and a blanket of air called the atmosphere, the earth would just be a ball of rock moving around the sun. Our planet has seas and an atmosphere because it is big enough for its gravity to keep them from flying away into space. Earth is also just the right temperature for water to exist as a liquid.

■ WHERE DID THE AIR COME FROM?

The air we have today, except for the oxygen in it, came mostly from early volcanoes. When earth first formed, it was covered with hydrogen. But hydrogen is a very light gas, and it soon leaked away into space.

The new air that soon began to cover the earth came from volcanoes. Their eruptions released water vapor and carbon dioxide from deep underground. This new atmosphere was thick and steamy, something like the air on Venus today. It had none of the oxygen that animals need to stay alive. If you had been there, you would have needed a space suit in order to breathe!

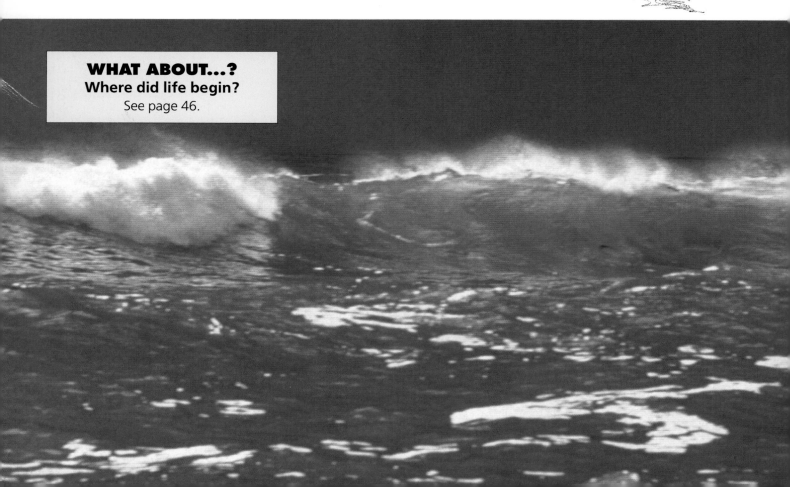

WHAT ABOUT...?
Where did life begin?
See page 46.

■ HAS THE SKY ALWAYS BEEN BLUE?

The earth's first atmosphere was like a thick fog. There would have been no blue sky for millions of years.

"White" sunlight contains all the colors of the rainbow. Air molecules are the right size to bounce, or scatter, blue and violet sunlight, but they let the other colors through. The sky is blue because blue light from the sun is bounced around from air molecule to air molecule until it comes at us from the whole sky.

Clouds look white because the droplets of water in clouds are bigger than air molecules and scatter every color in sunlight. On the early earth, the air was full of clouds, right down to ground level. In fact, the clouds held all the water that is now in the oceans. Until those oceans formed, every day was a foggy day. There was no blue sky to see.

■ WHY IS THERE OXYGEN IN OUR AIR?

There is oxygen in our air today because green plants put it there.

Whenever fire burns or metal rusts, oxygen is taken out of the air. Oxygen gets involved in chemical reactions, like burning and rusting, very easily. That is why there was no oxygen in the early atmosphere. Chemical reactions took it all out.

About 2 200 000 000 years ago — almost halfway through the story of the earth — that started to change. Life had already begun. Some living things had developed a way to make food by taking carbon from carbon dioxide, leaving oxygen that escaped into the air. This process, called photosynthesis, is what green plants do today. Without it, we would have no oxygen to breathe.

▼ HOW DID THE SEAS FORM?

After the new earth cooled, rain fell for hundreds of years. The first oceans probably covered the whole world.

At first, the new earth was covered with lava, and was too hot for liquid water. Instead, clouds covered the planet. Once the lava cooled and hardened, rain began to fall. It fell until the world was covered by a deep ocean.

By about 4 200 000 000 years ago, the sky had begun to clear a little. Now and then giant planetesimals still smashed into the earth. Some scientists think that these crashes may have built up enough heat to boil the ocean water away. Once the earth cooled, huge rainstorms would fill the ocean basins again.

PACIFIC OCEAN ALONG THE WEST COAST TRAIL, VANCOUVER ISLAND, BRITISH COLUMBIA, CANADA

Energy from the Sun

Our earth is bombarded by the sun's energy. Without it, the earth would be a dead place. Its water, and even its air, would be frozen solid.

Not all energy from the sun is good for living things, though. Ultraviolet rays can be dangerous. Luckily, the earth has a shield, the ozone layer, that protects us.

BE SURE TO WEAR A HAT IN THE SUN.

▼ WHY DO WE NEED THE OZONE LAYER?
The ozone layer protects us from the sun's harmful ultraviolet rays.

Two oxygen atoms joined together make an oxygen molecule. Three oxygen atoms joined together make ozone. There has been an ozone layer in our upper atmosphere for as long as there has been oxygen in our air.

Ultraviolet rays have more energy than rays of heat or light. We can handle a little ultraviolet energy. That is how lighter-skinned people get suntans. But too much can cause diseases like skin cancer. Too much can also harm the tiny plants in the sea that help cool the earth and provide food for many creatures. It's important to protect the ozone layer, because it protects us.

■ WHY DOESN'T THE EARTH GET TOO HOT?
The earth's atmosphere, oceans, and even its plants help keep it from becoming too hot or too cold.

The moon is as far from the sun as we are. But temperatures there rise to 137°C (279°F) and fall to −169°C (−273°F) — more than twice as hot or cold as it ever gets on earth. Why doesn't that happen here?

Unlike the earth, the moon has very little atmosphere. The earth's atmosphere helps to control our planet's temperature by soaking up some of the sun's energy, or reflecting it back into space. Water vapor and carbon dioxide in the atmosphere act like a blanket, keeping us from getting too cold.

Some scientists believe that, when it gets too warm, millions of tiny plants floating in the sea give off a chemical that helps clouds form. The clouds shade the surface, and the earth cools again.

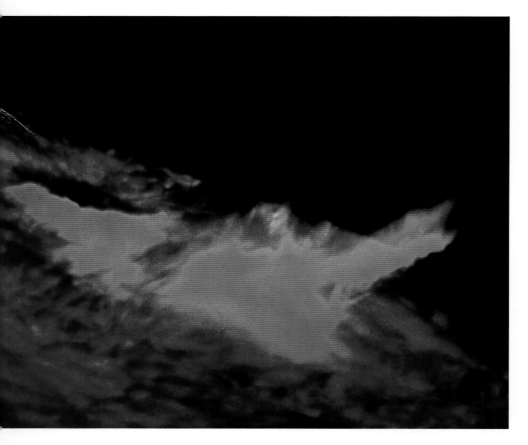

SOLAR FLARES LIKE THIS ONE PRODUCE BURSTS OF ULTRAVIOLET RADIATION.

▶ WHAT IS AN AURORA?
An aurora is a glow in the sky caused when particles from the sun collide with atoms in the upper air.

Storms on the sun may drive tiny, electrically charged particles called ions into space. Some of these ions travel toward the earth. When they get here, our planet acts as a giant magnet, and turns them toward the North or South Pole. There, they collide with atoms of oxygen, nitrogen and other gases at the top of our atmosphere. That makes the atoms give off light.

Each atom glows with a different color. Billions of glowing atoms make the beautiful displays we call auroras: the aurora borealis or northern lights in the Northern Hemisphere, and the aurora australis or southern lights in the Southern Hemisphere.

■ CAN AURORAS CAUSE PROBLEMS?
Auroras may be beautiful, but their electrical energy can break up radio, television or telephone signals. The glow from auroras has been mistaken for fires, or even UFOs!

Storms of electrical energy flowing to earth beneath an aurora can cause power blackouts, interfere with radio or television programs, trigger burglar alarms, or make automatic garage doors open by themselves.

Even the colors of an aurora can cause problems. Auroras are usually blue or green, but sometimes they are red. In 1938, a red aurora over England fooled firefighters into thinking that Windsor Castle was on fire.

AURORA BOREALIS, NORTHWEST TERRITORIES, CANADA

WHAT ABOUT...?
What is happening to the ozone layer?
See page 88.

Gravity from the Moon

No other planet in the solar system has a moon like ours. Some have moons that are bigger, like Jupiter's moon Ganymede, but Jupiter is so enormous that Ganymede seems tiny by comparison. Our moon is almost one-quarter the diameter of the earth. It is big enough to affect life here.

■ **WHY ARE THERE TIDES?**
The moon's gravity pulls the waters of the ocean, and of large lakes, toward it as the earth turns. The sun's gravity does too, but not as much, because the sun is much farther away.
The moon's gravity makes a "hill" of water on the side of the earth facing the moon. Strangely enough, it also makes another hill on the opposite side of the earth. This hill forms because the solid earth is pulled toward the moon faster than the water on the other side, and that water is left behind as a hill.

The hills of water move around the earth once a day as the planet revolves on its axis. Wherever one of the hills meets the shore, the water level rises. This is high tide. When the hill passes on, the water level falls until it reaches low tide. Because there are two hills, most places have two high tides and two low tides every day.

■ **WHAT DO SEASHORE ANIMALS DO WHEN THE TIDE GOES OUT?**
Many kinds of animals and plants spend all their lives in the intertidal zones along the seashore. At high tide they live underwater, but at low tide they have to be able to live above water without drying out.
On sandy beaches, animals like beach fleas can burrow deep in the sand at low tide to find water. On rocky shores, barnacles and mussels close their shells tightly at low tide to hold in water.

Even some fishes stay out of the water at low tide. Blennies hide in damp cracks in the rocks until the sea returns. On intertidal mudflats from East Africa to the South Pacific, mudskippers carry a supply of water in their gill chambers. They hop about like frogs on the exposed mud, searching for tiny crabs.

MUDSKIPPER
(*BOLEOPHTHALMUS* SP.)

HOW CAN I STUDY INTERTIDAL ANIMALS?
If you visit a rocky shore, look for tide pools at low tide. Tide pools are like little aquariums in the rocks, left behind when the tide goes out. Many intertidal animals move to the pools to stay wet at low tide. You may find snails, sea anemones, fishes and other animals there. But be sure to observe safety rules by the sea. Go with an adult, and wear rubber-soled shoes to climb on the rocks. And remember that the pools are the animals' homes. If you turn over any stones, be sure to put them back.

BAY OF FUNDY AT HIGH AND LOW TIDES, ALMA, NEW BRUNSWICK, CANADA

■ ARE ALL TIDES THE SAME?
High tides are higher when the moon is closer to the earth, or when it is in line with the sun.
The path of the moon around the earth is not a circle. Sometimes the moon is closer to the earth and sometimes farther away. When the moon is closest to the earth, high tides are 20 percent, or one-fifth, higher than when it is farthest away.

Since the sun pulls on the oceans, too, high tides are higher when the moon, the sun, and the earth are lined up and the moon and the sun are pulling together. These tides are called spring tides, because water "springs" up. When the moon's pull is at right angles to the sun's, high tides are lower. They are called neap tides.

▲ WHERE ARE THE HIGHEST TIDES IN THE WORLD?
The world's highest tides are in the Bay of Fundy, on the east coast of Canada.
When you slide back and forth in the tub, water sloshes higher if your slides take just about as long as it takes for the water to move from one end to the other. That is what happens in the Bay of Fundy.

Every 12½ hours, the North Atlantic tides push into the bay. It takes 13 hours for water to slosh from one end of the bay to the other. Because the bay is shaped like a narrow funnel, tides pile up even higher as they rush in. The water level at the head of the bay can rise more than 16 m (52 feet) from low to high tide.

WHAT ABOUT...?
What is a tidal wave?
See page 29.

The Dance of the Continents

When you look at a map of the world, do you ever notice that Africa and South America could fit together like pieces in a jigsaw puzzle? Could they once have been joined together?

In 1912, a scientist named Alfred Wegener said he thought that the continents drifted slowly over the earth. He believed that all the continents once came together in a single mass that he called Pangaea — Greek for "the whole world." Hardly anybody believed Wegener. But years later scientists discovered that he was right.

RED ANTARCTIC BEECH (*NOTHOFAGUS FUSCA*), NEW ZEALAND

DUCK-BILLED PLATYPUS (*ORNITHORHYNCHUS ANATINUS*), EASTERN AUSTRALIA

FOSSIL ANTARCTIC BEECH (*NOTHOFAGUS* SP.), SEYMOUR ISLAND, ANTARCTICA

▲ WHAT DID A FOSSIL TOOTH TELL US ABOUT THE CONTINENTS?

In 1992, scientists discovered a fossil platypus tooth in South America. Today the platypus lives only in Australia, but the 63-million-year-old tooth shows that platypuses once lived in South America, too.

The duck-billed platypus, and its cousins the echidnas, are the only mammals that lay eggs. They belong to a very old group of mammals. Up until 1992, though, no platypus fossils had been found anywhere but in Australia.

The exciting discovery of the platypus tooth is one of many that show that, long ago, South America and Australia were joined together, and animals like the platypus could walk from one to the other.

▲ HOW DO WE KNOW WHERE THE CONTINENTS HAVE BEEN?

Continents that were once joined together may share the same fossil animals and plants.

This is the fossil leaf of an Antarctic beech tree. Today, Antarctic beech grows in Australia, New Zealand and South America. But this fossil comes from Antarctica, where today there are no trees at all.

When Antarctic beech first appeared, about 66 million years ago, Antarctica, Australia, New Zealand and South America were part of a giant continent called Gondwanaland. Africa was once part of Gondwanaland, too, but it broke away much earlier, and there are no beech trees there. When the rest of Gondwanaland broke up, each part carried its own beech forests away — except for Antarctica, where the forests were buried under tonnes (tons) of ice.

▼ WHAT IS THE "SKIN" OF THE EARTH LIKE?

The earth is covered with a thin "skin" of solid rock, the crust. The crust is made of pieces that fit together like pieces of a jigsaw puzzle. These pieces are called tectonic plates.

The crust of the earth is about 6 km (4 miles) thick under the oceans, and about 32 km (20 miles) thick on the continents. That may seem like a lot, but if you could dig through the crust, you would still have to travel more than 6300 km (3800 miles) to reach the center of the earth.

Under the crust is the mantle, a layer of very hot, almost liquid rock. Beneath the mantle is the earth's core, a ball of iron and other metals. It is molten on the outside but has a solid center. The temperature of the core may reach 5000°C (9000°F).

The plates float on top of the mantle, carrying the continents with them.

■ HOW DO THE PLATES MOVE?

The liquid rock under the crust is moving. The movement carries the plates with it, like rafts on the sea.

You might wonder how the plates can move, since they all touch each other. When plates meet under the ocean, where the crust is thin, one starts to slide underneath the other. It starts to melt as it is pushed deeper into the mantle. When plates move away from each other, liquid rock from the mantle rises into the space between them, cools, and hardens, adding new crust to the plates.

On land the crust is too thick for the plates to slide over each other. There, the plates collide, crashing into each other like slow-moving bumper cars. As they hit, their edges may be thrust upward as mountain ranges. The Himalayas probably formed in this way.

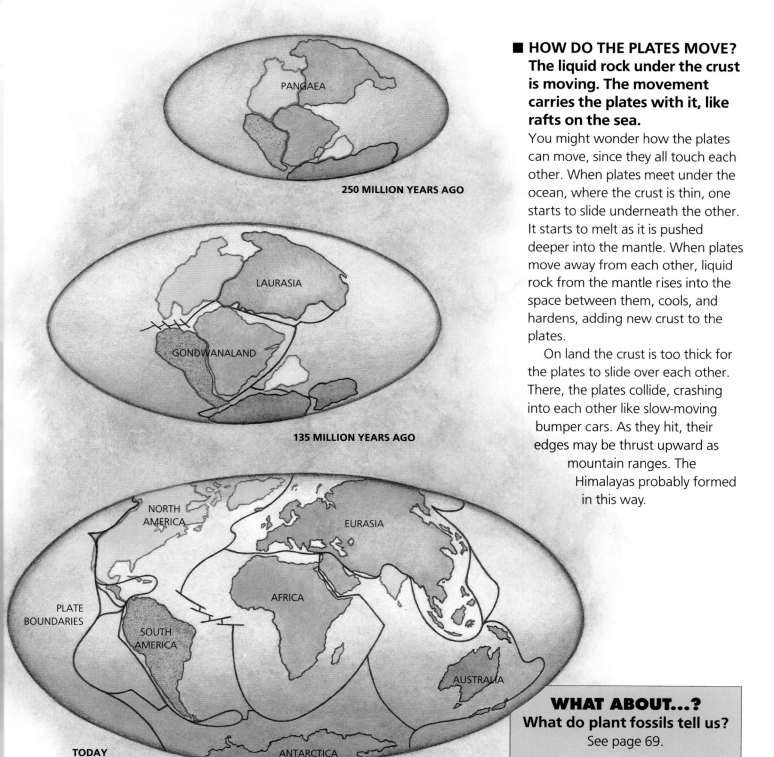

PANGAEA

250 MILLION YEARS AGO

LAURASIA

GONDWANALAND

135 MILLION YEARS AGO

NORTH AMERICA

EURASIA

PLATE BOUNDARIES

AFRICA

SOUTH AMERICA

AUSTRALIA

ANTARCTICA

TODAY

WHAT ABOUT...?
What do plant fossils tell us?
See page 69.

Where Do Rocks Come From?

There are many different kinds of rocks, but there are only three ways to make them. The first way is to take hot, liquid rock like lava and let it harden. This is like letting melted sugar harden to make toffee. The second way is to squeeze sand, or mud, or pebbles together. This is like pressing crispy rice cereal and melted marshmallows together to make squares. The third way is to take rocks made in one of the first two ways and change them into something new with heat and pressure. This is what you do with candy or crispy rice squares when you chew them.

▼ WHAT ROCKS HAVE FOSSILS IN THEM?

After millions of years, layers of sand, mud, or even small pebbles and shells can turn into rocks. These are called sedimentary rocks, and they are the only kinds of rocks that have fossils in them.

Rocks, shells and bits of coral are constantly being ground down into small pieces by wind and water. That is how sand is made. Layers of these small pieces, called sediments, can be "recycled" into new rock.

As more and more sediments pile up — perhaps washed onto a beach by the waves — their weight squeezes pieces in the deeper layers together. The remains of animals and plants in the sediments become fossils. Minerals such as quartz may form around the pieces, cementing them together like glue. The result is a sedimentary rock, such as sandstone, limestone or shale.

■ CAN ROCKS FLOAT?

Pumice is an igneous rock full of air bubbles. It is so light that it can float.

Have you ever said that something is as "heavy as a rock" or that it "sinks like a stone"? You obviously weren't talking about pumice.

Pumice is formed from molten lava that is full of bubbles of gas, like the froth on a glass of pop. It hardens so quickly when it reaches the cool air that the bubbles are frozen in place. The result is a piece of rock that is mostly air.

SEDIMENTARY ROCK, ALBERTA, CANADA

■ HOW CAN ONE KIND OF ROCK TURN INTO ANOTHER?

Heat and pressure can change one kind of rock into another. The new, "changed" rock is called metamorphic, from the Greek word for "change."

As new rocks form on top of old rocks, their weight presses down on the older layers. When the earth's crust moves, it may be bent or folded. This puts great strain on rocks. Magma rising into the earth's crust heats rocks around it, perhaps as hot as 700°C (1290°F). Differing combinations of heat, pressure and strain can change one kind of rock into another.

Shale is a sedimentary rock, but heat and pressure turn it into slate, a metamorphic rock sometimes used to make roof tiles. Granite is an igneous rock, but under very strong pressure it can become gneiss, a metamorphic rock that forms much of the earth's crust under the continents.

WHAT ABOUT...?
How are fossils formed?
See page 69.

▶ WHAT ROCKS START OUT AS MOLTEN LIQUID?
Igneous rocks begin as hot, molten liquid that cools and hardens.

"Igneous" comes from the Greek word for "fire." Igneous rocks start as molten rock, or magma, beneath the earth's crust. When magma cools, the minerals in it form crystals, the same way water does when it freezes into ice. The faster it cools, the smaller the crystals.

This lava flow cooled quickly after pouring out of a Hawaiian volcano. It has tiny crystals and looks smooth. Magma may also ooze upward through cracks in the earth's crust, where it may take a long time to cool — perhaps millions of years. The resulting rock will have big crystals in it. That is how granite, a hard rock used in many buildings, forms.

FRESH LAVA FLOW, HAWAII, USA

HOW CAN I START A ROCK COLLECTION?
Collecting rocks can be as easy as going outside and picking them up. But finding good specimens may be more difficult. Your local museum may run collecting trips, have a list of the best local places to visit, or give you help in identifying your finds. Remember to get permission before you collect rocks, and never take specimens from a national park. If you use tools like geologist's hammers, make sure you wear eye protection.

Remember that collecting doesn't stop when you get home. You should label your specimens so you know where they came from. Egg cartons make good specimen boxes for your collection. Look at the books in the reading list at the end of this book for more ideas.

BASALT COLUMNS, GIANT'S CAUSEWAY, ANTRIM, NORTHERN IRELAND

◀ WHAT IS THE GIANT'S CAUSEWAY?
The Giant's Causeway is a row of stone columns on the coast of Northern Ireland. People used to think that it was built by giants. It is really made of lava that flowed into the sea and cooled millions of years ago.

When molten lava flows into the sea, the water can cool it very quickly. When that happens the cooled lava, or basalt, sometimes breaks into six-sided columns.

Some of the Giant's Causeway's basalt columns are 6 m (20 feet) high and 50 cm (20 inches) across. According to local folk tales, the causeway was made by a race of giants who were trying to build a road to the nearby island of Staffa.

How to Build a Mountain

The surface of the earth is changing all the time. Usually the changes happen too slowly for us to see. But, over millions of years, these changes make mountains rise and fall. Today new mountains still grow, and old ones are still being worn down by wind and water.

MOUNT FUJI, JAPAN

■ HOW DOES THE EARTH BUILD MOUNTAINS?
Rock is bent or pushed up into mountains when the plates on the earth's crust collide.
Most of the mountain-building action on earth happens at the edges of the great tectonic plates (see page 21). Imagine two cars crashing into each other. The force of the collision can make the metal of the cars buckle into folds, or even break. Leaking fuel can catch fire.

The same thing happens when two plates collide. The rocks can be bent into folds, forming fold mountains. Great cracks called faults can form where the rocks break. Blocks of rock between the faults can be thrust upward into block mountains. And molten rock from within the earth can explode upward into volcanoes. The Cascades of the western United States, for example, are a mixture of block mountains and volcanoes.

▲ WHAT IS THE "RING OF FIRE"?
There are so many volcanoes around the rim of the Pacific Ocean that it is called the "Ring of Fire."
Mount Fuji in Japan, Mount Saint Helens in the United States, and Mount Pinatubo in the Philippines are only a few of the hundreds of volcanoes in the ring. They are there because the floor of the Pacific is spreading. Its edges are being pushed deep into the earth under the continents. There the edges melt, forming great pools of molten rock that fuel the volcanoes.

► HOW MANY TIMES HAVE THE ROCKY MOUNTAINS FORMED?
The Rocky Mountains of western North America have built up three times.
About 320 million years ago, the first Rocky Mountains were pushed up. Over millions of years, they were worn down again. About 70 million years ago, new mountains formed and were also worn away. Thirty-five million years ago, huge volcanoes erupted, and 26 million years ago the land was lifted again. Today's Rocky Mountains are the third range to be built on the same spot!

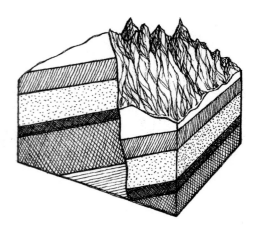

FOLD MOUNTAIN (LEFT); BLOCK MOUNTAIN (RIGHT)

ROCKY MOUNTAINS, BANFF NATIONAL PARK, ALBERTA, CANADA

WHAT ABOUT...?
How do rocks move along a fault?
See page 29.

Mountains of Fire

Volcanic eruptions may destroy mountains, create huge "tidal waves," or tsunamis, and bury cities. They may change the world's weather by throwing so much dust into the air that it blocks some of the sun's heat, as the Philippines' Mount Pinatubo did in 1992.

But volcanoes are also builders. They have raised whole island chains out of the sea, like Hawaii and the Galapagos. Mount Kilauea's spectacular eruptions are still adding rock to the big island of Hawaii.

▲ WHY DO VOLCANOES ERUPT?
Volcanoes erupt when hot liquid rock from inside the earth rises to the surface. Gases in the rock boil away, causing an explosion.
Molten rock, or magma, is lighter than solid rock. It floats slowly upward like a cork in syrup. When the magma gets near the surface, gases in it boil away. This creates an explosion that throws magma to the surface as lava, along with ash and chunks of rock.

The more gas there is, and the more syrupy the magma, the more violent the eruption. This is because it is harder for the gas to escape from thicker magma than from thinner magma. When it finally does, it explodes with greater force.

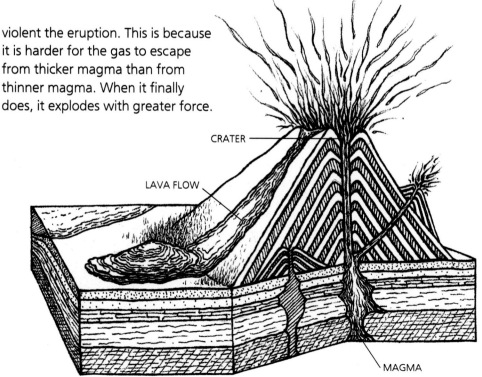

CRATER

LAVA FLOW

MAGMA

WHAT ABOUT...?
What is the "Ring of Fire"?
See page 24.

MOUNT KILAUEA IN ERUPTION, HAWAII, USA

■ HOW DO VOLCANOES BUILD THE LAND?

When lava cools, it forms solid rock. Later eruptions add more rock. The buildup of lava, ash and boulders forms the volcano. When a volcano under the sea builds up high enough, it becomes an island.

The Hawaiian Islands were formed by a "hot spot" — a large plume of magma under the floor of the Pacific Ocean. Volcanoes erupting over the hot spot built up large islands. The hot spot doesn't move. But over the centuries the floor of the ocean moves, carrying the islands away. New ones form in their place. The result is a long island chain.

However, lava rock is soft and easily worn down. Once away from the hot spot, with no fresh lava available, the islands begin to shrink as the rain and wind wear the lava into dust. As you travel to the northwest, the Hawaiian Islands get smaller and smaller, until nothing is left but undersea mounds.

■ WHAT WERE THE BIGGEST ERUPTIONS IN HISTORY?

Big volcanic eruptions don't happen very often. When they do, they may change the course of history.

Santorini (Thera), Greece, around 1628 BC: A huge eruption turned most of this island into dust. This eruption may have destroyed the Minoan civilization. It also probably started the story of Atlantis, a land that was supposed to have vanished beneath the sea on a single day.

Mount Vesuvius, Italy, AD 79: This famous eruption buried the Roman cities of Pompeii and Herculaneum.

Mount Tambora, Sumbawa, Indonesia, AD 1815: Volcanic dust from this eruption, the largest in history, circled the earth. It blocked out so much of the sun's heat that the next year, 1816, was called "the year without a summer."

Krakatau (or Krakatoa), Indonesia, AD 1883: This island was practically destroyed in a blast heard 4000 km (2480 miles) away.

Earthquakes

If you don't live in an earthquake zone, the idea of the earth shaking may seem very strange to you. But earthquakes happen all the time. There are more than a million every year, though most are too weak to feel.

Earthquakes happen along faults, which are cracks in the earth's crust where the rocks on either side have moved. When the plates of the earth's crust move, they push rocks together or pull them apart along fault lines. But the rocks don't move easily. A lot of pressure must build up before they do. The release of energy when the rocks finally move creates the shock wave of the earthquake.

EARTHQUAKE DAMAGE IN SAN FRANCISCO, CALIFORNIA, USA, OCTOBER 1989; THE QUAKE MEASURED 6.9 ON THE RICHTER SCALE.

SHOULD I WORRY ABOUT EARTHQUAKES?

If you don't live in an earthquake zone along a major fault line, you may never feel an earthquake. If you do live in such an area, you have probably already felt dozens of them. Most quakes are too weak to cause any damage, and won't shake you up as much as a drive over a bumpy road. Today, the buildings in most cities in earthquake zones have been specially designed to cushion the shock of a quake.

▲ HOW DO WE MEASURE EARTHQUAKES?
We can measure how strong an earthquake is with a seismometer.

When the rocks finally move along a fault, shock waves travel round the world through the earth's crust, sometimes more than once. Other waves pass right through the center of the earth. Seismometers measure the energy in these waves. The more energy that built up in the rocks before the quake, the stronger the shock wave.

We rate earthquakes on the Richter scale. Each number on this scale is ten times stronger than the one before — a five-point earthquake is ten times stronger than a four-point, a six-point ten times stronger than a five-point, and so on. A four-point earthquake might just shake the ground a little, but a six-pointer could do a lot of damage to buildings, bridges and streets. The very worst quakes measure about 8.5 on the Richter scale.

◄ HOW DO ROCKS MOVE ALONG A FAULT?
Rocks along a fault may move sideways, away from each other, or toward each other.
The famous San Andreas Fault in California, USA, marks the line where the Pacific plate is slowly moving north past the North American plate. The rocks on either side grind past each other like gears someone has forgotten to oil. In 1906, an earthquake along the San Andreas Fault moved roads and fences sideways by as much as 6 m (20 feet).

Where the crust is being stretched apart, one block of land may slide downhill past another. This rarely causes serious quakes. But if the rocks are being squeezed together, one side may be pushed "uphill" against the other. This requires a lot of energy, and can cause very damaging quakes.

EARTHQUAKE DAMAGE ALONG THE SAN ANDREAS FAULT, CALIFORNIA, USA

WHAT ABOUT...?
How do the plates of the earth's crust move?
See page 21.

SAND AND MUD BROUGHT UP BY AN EARTHQUAKE, SAN FRANCISCO, CALIFORNIA, USA

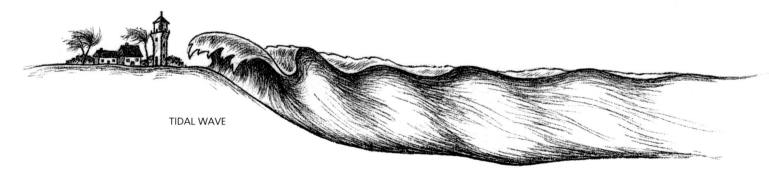

TIDAL WAVE

▲ WHAT IS A TIDAL WAVE?
A tidal wave, or tsunami, has nothing to do with tides. It is a giant wave caused by an undersea earthquake.
If you shake or tap a basin of water, you send a shock wave through it that can make little waves on the water's surface. Do it hard enough, and the water may start to slosh against the sides of the basin.

An undersea earthquake or volcanic eruption can do that to the whole ocean. The shock waves it sends through the ocean basin can create the monster wave we call a tsunami, or tidal wave. Tidal waves can move at 800 km/h (500 mph) at sea. Near shore they slow down but rise higher and higher. They can be 50 m (160 feet) high when they hit the coast, high enough to engulf a ten-story building.

WHERE CAN I SEE METEORITES?

Any piece of a meteor that has fallen to earth is called a meteorite. Some are mostly nickel and iron. Others are made of stone. The largest one ever discovered is on display at the American Museum of Natural History in New York City. It weighs 31 000 kg (34 tons). You can probably see smaller ones in your local natural history museum. One way you almost certainly won't see a meteorite is by having one fall on you. That has happened only once in historic times, to Ann Hodges of Sylacauga, Alabama, USA. On November 30, 1954, a 4-kg (9-pound) meteorite crashed through the roof of her house, bruising her on the hip.

Visitors from Space

The moon has no wind or water to wear down its surface. Instead, it is covered with impact craters. The craters were made by meteors — rocks that fell from space. Most of them fell 4 million years ago, when the solar system was still full of the rocks that formed the planets.

A rain of meteors fell on earth, too, in those early days, but the craters they made have long since disappeared. They were destroyed as the surface of the earth changed. Every few million years, though, a giant meteor still smashes into the earth. Some may have changed the history of life itself.

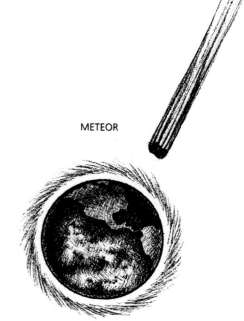

METEOR

◄ HOW DID METEOR CRATER FORM?
Meteor Crater is in Arizona, in the United States. It was created about 49 500 years ago by the crash of a meteor that weighed 50 000 000 tonnes (tons).
Meteor Crater is the best-preserved of the 200 or so impact craters known on earth, and the first one proven to have been made by a meteor. It is 1.2 km (¾ mile) across and 174 m (570 feet) deep.

The meteor that made it was a huge lump of nickel and iron almost 30 m (100 feet) across. It struck the earth at a speed of 69 000 km/h (43 000 mph). It hit so hard that most of it was smashed into dust. The force of the blast heaved rocks up around the edge of the crater. That is why the crater is not just a hole in the ground, but a hole with a circular hill around it.

■ DID A METEOR KILL THE DINOSAURS?
A huge meteor probably hit the earth around the time the dinosaurs became extinct. We still don't know if it killed them.
In 1980, Luis Alvarez and his son Walter announced their discovery that rocks from the time the dinosaurs died were full of iridium. Iridium is rare on earth, but common in meteorites. The iridium layer suggested that a giant meteor had struck the earth, scattering bits of itself over the whole world. Some scientists think they have found where it hit, in Mexico's Yucatan Peninsula.

Clouds of dust from the crash probably circled the world, blocking out the light of the sun. The cold and dark may have killed so many plants that the dinosaurs died of starvation. However, many scientists think that the dinosaurs were dying out anyway, long before the meteorite hit.

■ WHAT IS A SHOOTING STAR?
A shooting star is the trail a falling meteor makes as it burns up in the earth's atmosphere.
Most meteors never make it to the ground. Friction as the falling meteor collides with air molecules in the earth's atmosphere makes them so hot that they start to burn. Because almost all meteors are very small, they usually burn up completely high in the air.

You can see shooting stars on almost any clear night. At certain times of the year, though, the earth passes through clouds of meteors that circle the sun. This causes a meteor shower, when there are far more shooting stars than usual. During the Perseid meteor shower, which happens around August 12, you may see as many as one shooting star per minute.

WHAT ABOUT...?
What was there before the earth formed?
See page 10.

METEOR CRATER, ARIZONA, USA

Water Shapes the Land

Every breaking wave and every flowing river wears away the surface of the earth, bit by tiny bit. Even a fall of rain can wash soil away, or seep below ground where it dissolves minerals in the rocks. Moving water is the most important of all the forces of erosion.

CAVE CRICKET (*CEUTHOPHILUS* SP.), NORTH AMERICA

▶ **WHERE DID THE GRAND CANYON COME FROM? The Grand Canyon was dug by the Colorado River. It is 446 km (277 miles) long and 1800 m (6000 feet) deep.**
Every river digs bits of the earth away. The faster it flows, the more it digs, because the water hits the ground with more force. A slow-flowing river may dig a shallow bed. A fast-flowing river may carve a deep valley. The biggest river valley in the world is the Grand Canyon in the southwestern United States.

While the Colorado River was digging its valley, the land around it was rising. The higher the land rose, the faster the river flowed and the deeper the canyon became. The river has cut its way down to rocks that are almost 2 000 000 000 years old, and is still digging.

GRAND CANYON, ARIZONA, USA

STALACTITES, MAMMOTH CAVE, WESTERN AUSTRALIA

WHAT ABOUT...?
Where are the biggest lakes
and rivers?
See page 37.

◄ HOW DO CAVES FORM?
Most caves are dug by underground water. The water dissolves minerals in the rocks and carries them away. The space that is left is a cave.

Not all caves are dug by water. Some are made by flowing lava. On Mount Elgon in Kenya there are caves that were probably dug by elephants looking for salt. But the most spectacular caves, like Mammoth Cave in Australia, have been slowly carved in limestone by rainwater.

Rain and carbon dioxide from the air and soil make a very weak acid that eats away at limestone. It seeps underground, where after millions of years it can form huge caves.

In the caves, the water leaves behind its dissolved minerals. These make strange formations, such as the hanging icicles of stone, called stalactites, and the upside-down stone icicles growing up from the cave floor, called stalagmites.

■ WHERE ARE THE WORLD'S BIGGEST CAVES?
The world's longest cave systems are the Mammoth and Flint Ridge caves in Kentucky, USA. Together they contain 530 km (329 miles) of underground passageways.

The largest known single underground chamber in the world is Sarawak Chamber, in Gunung Mulu National Park on the island of Borneo. It is 700 m (2297 feet) long and 400 m (1312 feet) wide — big enough to hold eight jumbo jets. The world's deepest known cave is Gouffre Jean Bernard in France. It lies 1535 m (5036 feet) underground.

Cave formations can be very large, too. In the Cueva de Nerja in Spain, for example, there is a stalactite that is 59 m (195 feet) long — longer than New York City's Statue of Liberty is tall.

▶ CAN THE OCEAN CHANGE THE LAND?
Ocean waves can polish stones and grind them into sand, carve cliffs, and even dig caves along the shore.

Behind this Nova Scotia beach are basalt columns like the ones on the Giant's Causeway (see page 23). The smooth, round stones on the beach were once rough pieces of rock broken away from the columns by the pounding waves. They got their shape as waves rolled them back and forth. They ground against each other, losing their rough edges. Grit and sand polished them smooth.

COAST OF BRIER ISLAND, NOVA SCOTIA, CANADA

Rivers of Ice

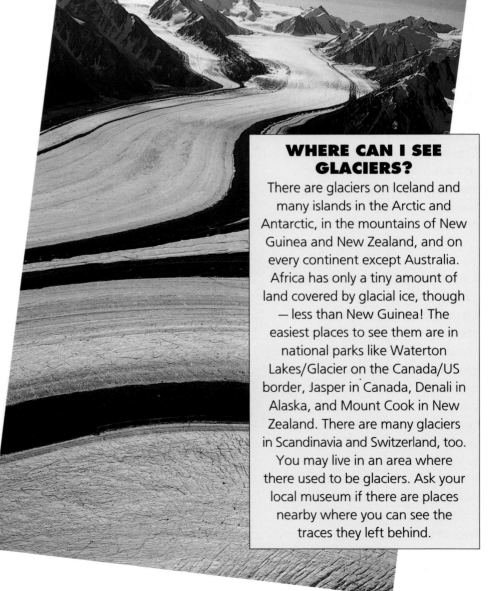

ICEBERG

Only 12 000 years ago — not long when you remember how old the earth is — much of our world was covered by thick sheets of ice. We call that time the Ice Age. But the Ice Age is not really over. Giant ice caps still cover most of Antarctica and Greenland. Alpine glaciers — slow-moving rivers of ice — fill mountain valleys around the world. Even where the glaciers have disappeared, we can see signs of their passage: beautiful mountain valleys; hills of earth, or moraines, pushed aside by tonnes (tons) of ice; and huge boulders carried a long way from where they were formed.

► HOW ARE GLACIERS FORMED?

If snow does not melt during the summer, it piles up year after year. The snow becomes ice, and the ice may form a glacier.

There does not have to be a lot of snow for a glacier to form. It hardly ever snows in central Antarctica, for example. The important thing is that some of the snow does not melt in the summer. That happens near the North and South poles, or on the tops of mountains. The weight of the snow squeezes the flakes together into ice. The ice moves slowly down the mountain valleys like a frozen river until it reaches a point where it melts or breaks up.

MEETING PLACE OF THE SOUTH AND NORTH KASKAWULSH GLACIERS, YUKON TERRITORY, CANADA

WHERE CAN I SEE GLACIERS?

There are glaciers on Iceland and many islands in the Arctic and Antarctic, in the mountains of New Guinea and New Zealand, and on every continent except Australia. Africa has only a tiny amount of land covered by glacial ice, though — less than New Guinea! The easiest places to see them are in national parks like Waterton Lakes/Glacier on the Canada/US border, Jasper in Canada, Denali in Alaska, and Mount Cook in New Zealand. There are many glaciers in Scandinavia and Switzerland, too. You may live in an area where there used to be glaciers. Ask your local museum if there are places nearby where you can see the traces they left behind.

WHAT ABOUT...?
How does global warming happen?
See page 88.

ICEBERG OFF BAFFIN ISLAND, NORTHWEST TERRITORIES, CANADA

HOPE BAY, ANTARCTICA, WITH ADELIE PENGUINS (*PYGOSCELIS ADELIAE*)

▲ WHAT IS AN ICEBERG?
An iceberg is a piece of a glacier that has broken off where the glacier meets the sea.

In the far north or far south, the surface of the sea may freeze. This sea ice, though, is very different from the ice in icebergs. Icebergs are made of freshwater ice. Some people have even suggested towing icebergs south to California, to provide fresh water for cities like Los Angeles. Nobody has actually tried it yet.

The biggest icebergs come from the edges of the ice sheets of Greenland and Antarctica. In Antarctica, the ice sheet actually reaches out over the Ross Sea itself, and huge pieces may break off it. One seen in 1956 was 335 km (208 miles) long and 97 km (60 miles) wide — the size of Belgium. Icebergs are very dangerous for ships, because only one-eighth of the berg shows above the water.

■ HOW FAR DO ICEBERGS TRAVEL?
An iceberg from Antarctica once drifted north for 5500 km (3440 miles), almost into the tropical Atlantic near Rio de Janeiro, Brazil.

Antarctic icebergs drift north on ocean currents at around 13 km (8 miles) a day. Because icebergs may take years to melt, they can drift a long way. Usually they don't get past the cold waters of the sub-Antarctic. Sometimes, though, a storm will push one much farther north.

Arctic icebergs may drift a long way, too — far enough south to be a danger to ships. One of the most famous shipwrecks in history was the sinking of the *Titanic* in 1912. The ship struck an iceberg on its way from England to the United States.

■ WHAT WOULD HAPPEN IF THE ICE CAPS MELTED?
If all the ice in Antarctica melted, the oceans would rise as much as 65 m (210 feet), high enough to cover a fifteen-story building.

The Antarctic ice cap holds nine-tenths of the world's ice. It may have taken 100 000 years of snow to make it. At one point it is 4776 m (15 670 feet) thick. That is higher than most mountains. The ice is so heavy that it actually pushes the land of Antarctica down, flattening the whole planet at the South Pole.

It might take centuries for all that ice to melt once it started. But scientists are worried that the world may be getting warmer, and that enough ice might melt to raise the level of the sea by a meter (3 feet) or so. That would be enough to flood many of our coastal cities.

Fresh Water

WILD HORSES IN THE CAMARGUE (DELTA OF THE RHONE RIVER), FRANCE

Our earth has plenty of water. Most of it is salt water, though. And most of the remaining fresh water is frozen in the ice caps of Antarctica and Greenland. The fresh water we have — whether it falls as rain, flows in a river, seeps underground, or sits in a lake or a marsh — is precious, not just to us, but to all life on land.

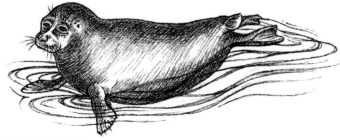

BAIKAL SEAL (*PHOCA SIBIRICA*), LAKE BAIKAL, RUSSIA

■ **HOW DO LAKES FORM?**
To make a lake, you need a hollow in the ground, water flowing into it, and something to keep most of the water from flowing out again.
Lakes can fill hollows made by glaciers, like the Great Lakes in North America. Lakes can fill volcanic craters, like Crater Lake in Oregon, USA, or places where the land has sunk between two faults, like the Rift Valley lakes of east Africa.

Not all lakes are filled with fresh water. Rivers flowing into a lake carry bits of salt. If none of the water in a lake can escape, except by evaporating into the air, it will get saltier and saltier. That's why the ocean is salty. Some salt lakes, like the Dead Sea, are even saltier than the ocean.

WHAT ABOUT...?
What is acid rain?
See page 87.

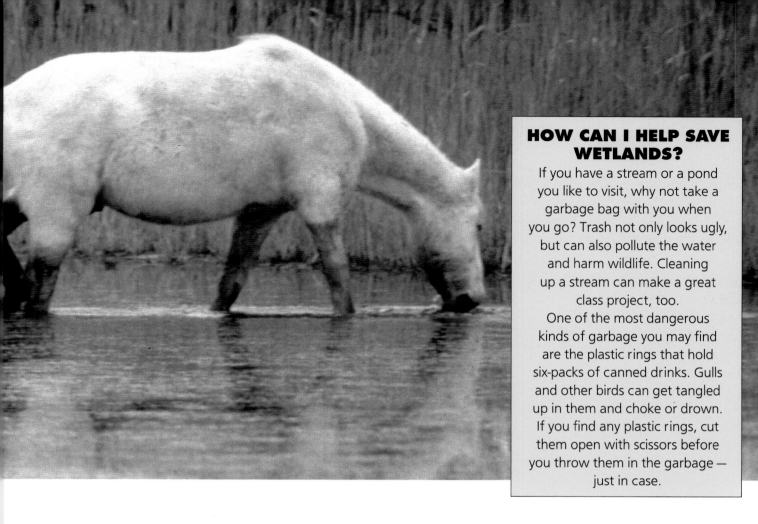

HOW CAN I HELP SAVE WETLANDS?

If you have a stream or a pond you like to visit, why not take a garbage bag with you when you go? Trash not only looks ugly, but can also pollute the water and harm wildlife. Cleaning up a stream can make a great class project, too.

One of the most dangerous kinds of garbage you may find are the plastic rings that hold six-packs of canned drinks. Gulls and other birds can get tangled up in them and choke or drown. If you find any plastic rings, cut them open with scissors before you throw them in the garbage — just in case.

■ WHERE CAN YOU FIND A LAKE WITHOUT WATER?

Lake Eyre, in the central desert of Australia, may fill with water only once every thirty years or so.

Lake Eyre is Australia's largest lake, but it is usually a lake without water. Instead of water, its bed is covered with 6000 km² (2400 square miles) of dry salt. Only a few lizards and ants live on it.

Sometimes — perhaps only two or three times a century — enormous rainstorms fall in the desert, and Lake Eyre and other dry lakes nearby fill with water. Soon they swarm with shrimps, fishes and frogs that have lain dormant underground or have been washed into the lakes from desert pools. Thousands of water birds flock from all over Australia to nest on their shores. But within a year or so, the lakes are dry again.

▲ WHERE DO SWAMPS COME FROM?

A lake, or a river delta, may gradually fill up with sand and soil, until nothing is left but a swamp.

The water flowing into lakes carries sand and dirt that can fill them in. As plants grow in the shallow water, the lake becomes a swamp or marsh. A river delta, where a river drops its soil before entering the sea, also forms marshes like the Camargue in southern France.

A marsh can be one of the best places to find wildlife. The Camargue teems with birds and fish. Thousands of ducks from all over Europe winter there, making the marsh important not only for France, but for the wildlife of many other countries, too.

■ WHERE ARE THE BIGGEST LAKES AND RIVERS?

Three-quarters of the lake water in the world is in one lake, Siberia's Lake Baikal. One-fifth of the river water is in one river, the Amazon.

Lake Baikal is the deepest lake in the world. It is 1.7 km (1 mile) deep at its deepest point. More than 950 kinds of animals, including a type of seal, live nowhere else. The shallow, salty Caspian Sea between Europe and Asia is the world's biggest lake by area; the biggest freshwater lake by area is Lake Superior in North America.

The world's largest river is the Amazon in South America. It is 6400 km (4000 miles) long and its mouth is 320 km (200 miles) wide. The water from 1100 other rivers flows into it. The Nile in Africa is just as long, but carries much less water.

Wet Lands and Dry

On one side of a mountain range, you may find a rainforest. Cross the mountains, and you may find a desert instead. Of all the things that make one part of the world different from another, water is the most important — but not just the water in rivers, lakes or the sea. The water vapor in the air makes the difference. How much the air carries, where the wind takes it, and where it falls as rain determines whether the land below is lush or barren.

■ **WHERE ARE THE WETTEST AND DRIEST PLACES ON EARTH?**
Earth's wettest spot is on the island of Kauai in Hawaii. Its driest is in Chile.
The driest place in the world is in South America, in the Atacama Desert, where it may only rain once every ten years. Another one of the world's driest places may surprise you — the South Pole. Even though it is covered with ice, the central plateau of Antarctica is a desert, where snow rarely falls.

The wettest place in the world is probably in Hawaii. Mount Waialeale, on the island of Kauai, receives 1140 cm (450 inches) of rain a year — enough to drown a three-story building. It rains there almost every day. Cherrapunji, India, had the rainiest single year ever recorded: 2646 cm (1042 inches) of rain fell between August 1860 and July 1861.

▼ **HOW DO RAINFORESTS COOL THE WORLD?**
Rainforest plants take up rainwater through their roots and recycle most of it back into the air through their leaves. Winds carry it away to fall as rain somewhere else.
Rainforests grow where rain falls almost every day. Rainforest plants recycle three-quarters of the rain back into the air as water vapor within forty-eight hours. This moist air circles north and south, bringing cooling rains to the temperate parts of the world.

When rainforests are destroyed, bare red earth may be left in their place. In the hot tropical sun, it heats up like a frying pan, drying the air above it. Instead of cool moist air, hot dry winds blow into the temperate zone. That is one reason why it is important to save the rainforests.

DOUBLE-CRESTED BASILISK (*BASILISCUS PLUMIFRONS*), RAINFORESTS OF CENTRAL AMERICA

ATLANTIC RAINFOREST, CARAGUATA RESERVE, SANTA CATARINA, BRAZIL

WHAT ABOUT...?
Why are the deserts growing?
See page 90.

SAND DUNE, DEATH VALLEY, CALIFORNIA, USA

▲ WHY ARE THERE DESERTS?
A desert is a place where less than 25 cm (10 inches) of rain falls a year.
Deserts like the Sahara in North Africa, the largest in the world, lie about 30° north and south of the equator. Here, dry air circling north and south from the equator spills back to earth, drying the land.

Deserts can form in other ways, too. Winds that pick up moisture from the ocean usually let it fall as rain over the land. But if a mountain range gets in the way, the wind may be forced higher and higher, where its air becomes cold. Cold air doesn't hold much moisture, and so the rest of the rain falls on the mountains. By the time the wind reaches the other side, it is too dry to water the land. The result is a desert.

■ ARE ALL DESERTS HOT AND SANDY?
Only about one-quarter of the world's deserts are covered with sand. Some deserts are cold, especially at night.
Most people think that all deserts are sandy, hot and barren. But many deserts in the world, like the Great Stony Desert in Australia, are covered with rocks and pebbles instead of sand. Most deserts are cold at night, because there is no moisture in the air to trap heat escaping from the ground. Some, like the Taklamakan Desert in China, are cold both day and night.

The Atacama Desert in Chile and Peru is truly deserted. Almost nothing lives there. But deserts like the Kalahari in Africa and the Mojave in North America are home to a wide range of fascinating animals and plants, from spiny lizards and burrowing mice to, in America, giant cacti, and, in the Kalahari, elephants.

SHORT-HORNED LIZARD
(*PHRYNOSOMA DOUGLASSI*),
DRY AREAS OF WESTERN
NORTH AMERICA

Storms and Seasons

The weather changes from day to day, or from season to season, but it stays much the same from year to year in any one place. That is what we mean by climate.

The climate and the weather are part of a worldwide system of moving masses of warm or cool air. The spinning earth swirls these circling air masses sideways, making them collide and mix together. It is this changing mixture of air that gives us different weather every day.

SNOWFLAKES

HAILSTONES

▲ **WHAT IS HAIL?**
Hailstones are balls of ice that grow inside thunderclouds. They can be as small as a pea, or as big as a grapefruit.
Because the temperature of the air in a thundercloud can be below freezing, the moisture in the cloud may freeze into pellets of ice. As the pellets start to fall, powerful winds blowing up through the cloud may carry them back up again, where more ice forms around them. This can happen over and over, until the pellet is too heavy to stay in the cloud. At that point, it has become a hailstone.

If you cut a hailstone open, you can see rings inside it, like those in an onion. A hailstone's rings show the layers of ice it grows as it forms.

■ **WHY ARE THERE SEASONS?**
There are seasons because the half of the earth that tips away from the sun has winter, while the half tipped toward it has summer.
If you could put the earth and the sun on a table, the earth would not stand straight up and down but would be tilted at an angle of 23.5°. As it orbits the sun, sometimes the "top" half (the Northern Hemisphere) is tilted toward the sun, sometimes the "bottom" half (the Southern Hemisphere).

Because the half tilted toward the sun receives more overhead sunlight and longer days, it is warmer than the other half. That is why it is summer in Canada when it is winter in Australia. It is also why, unless you live in the tropics or near the poles, there are four seasons. When the tilt is neither toward nor away from the sun, it is spring or fall.

WHAT ABOUT...?
What is an aurora?
See page 17.

▶ WHY DO STORMS HAPPEN?
Storms may happen when warm air and cool air collide. The cold air forces the warm air high into the sky, where its moisture falls as rain.

Warm air is lighter and can hold more water vapor than cold air. When the two meet, the warm air rises and cools. After it cools, it can't hold as much water vapor. The extra vapor forms clouds and rain.

In temperate climates, between the Arctic or Antarctic and the tropics, warm and cold air masses meet along a line called a front. If the warm air is moving toward the cold along a warm front, it slides slowly up over the cold, bringing cloudy, rainy days. If cold air is moving forward along a cold front, it pushes the warm air up quickly. This often brings thunderstorms.

STORM OVER TORONTO, ONTARIO, CANADA

◀ WHAT IS LIGHTNING?
Lightning is a giant electric spark running between a thundercloud and the ground, or from one cloud to another.

If you touch someone after shuffling your feet on the carpet, you may make a spark of static electricity. Lightning is exactly the same thing, but on a much larger scale.

A lightning bolt heats the air around it to five times the temperature of the surface of the sun — but only for a few millionths of a second. The heated air expands, then collapses back as it cools, making the noise we call thunder. Because sound travels more slowly than light, we hear the thunder after we see the lightning. You can tell how far away the lightning is by counting the seconds between the lightning and the thunder: three seconds for each kilometer, or five seconds for each mile.

THUNDERSTORM OVER TUCSON, ARIZONA, USA

Hurricanes and Tornadoes

They're called hurricanes in North America, typhoons in China and Japan, baguios in the Philippines and cyclones in Australia. Whatever their name, they are the largest storms in the world. Tornadoes are much smaller than hurricanes, but their winds may be even more powerful.

Hurricanes are very different from tornadoes. Hurricanes arise over the ocean, while the worst tornadoes form over land. A hurricane is a huge, spinning storm that may be hundreds of kilometers (miles) across. A tornado is only part of a storm, and may be no wider than a single house.

▼ HOW ARE TORNADOES BORN?
Tornadoes begin as part of a spinning thunderstorm.

Three-quarters of the world's tornadoes are born in the central United States, in places like Kansas (where Dorothy lived before she was blown to Oz, in L. Frank Baum's famous story). When conditions are right, warm moist air blowing in from the Gulf of Mexico may rise into the sky. Winds catch it, spin it over and over, and tip it on end. Higher up, the moisture in this spinning column forms a thunder-cloud.

In this kind of spinning storm, warm, dry air sometimes blows into the middle of the cloud. Here it cools and sinks quickly, drawing the spinning winds to the ground. The result is a tornado.

■ WHAT IS A TORNADO LIKE?
A tornado is a spinning funnel. Its winds may reach 480 km/h (300 mph) — the fastest winds known at ground level.

Some tornadoes are fairly short, but others can be 300 m (1000 feet) long. They may be only a few meters (feet) across, or as much as 3.2 km (2 miles) wide. Tornadoes sweep along the ground like huge vacuum cleaners. When they form over water, they are called water-spouts.

Tornadoes are not the only funnels of air. Harmless dust devils or whirly-whirlies often form in the desert, with no sign of a storm to start them off. Snow devils may spin up a flurry of snow, then disappear a few seconds later.

TORNADO

WHAT ABOUT...?
Why do storms happen?
See page 41.

HURRICANE ELENA, SEPTEMBER 1985,
PHOTOGRAPHED FROM A SATELLITE

▲ HOW IS A HURRICANE BORN?
A hurricane may develop if the waters of the ocean's surface are warmer than 27°C (80°F), and the conditions in the air are just right.

When the water is that warm, a lot evaporates into the air. This makes warm, humid air that rises. If the air is already disturbed, perhaps by rain showers, it may rush in to replace the rising air, causing winds that come together and blow upward. Winds start to spin around each other as the storm picks up strength. The result is a tropical storm. If the wind speed reaches 120 km/h (74 mph) or higher, the storm is called a hurricane.

As long as the storm stays over the warm ocean, more water vapor can rise from the sea to "fuel" it. Once it hits land or cold water, this "fuel" is no longer available and the storm usually dies away.

SHOULD I WORRY ABOUT HURRICANES OR TORNADOES?

If you don't live in a hurricane zone or a tornado belt, you probably have nothing to worry about. If you do live in such an area, remember that many countries have special weather departments to watch out for bad storms. They will warn your parents or teachers in plenty of time for you to take cover or get out of the way. For more information on hurricanes you can write the National Hurricane Center, 1320 South Dixie Highway, Coral Gables, Florida, USA 33146. For tornado facts, write the National Severe Storms Forecast Center, Room 1728, 601 East 12th Street, Kansas City, Missouri, USA 64106.

DAMAGE FROM HURRICANE GILBERT, PUERTO MORELOS, MEXICO

■ WHAT IS A HURRICANE LIKE?
A hurricane is a spinning pinwheel of wind and cloud with a calm, cloudless center called the "eye."

The strongest winds in a hurricane are the ones spinning around the eye in the "eye wall." They may reach 320 km/h (200 mph). Here, a ring of warm, humid air rises from the sea. When it cools, some of it spills inward and sinks slowly through the center of the storm. That creates the calm "eye." The eye can be 40 km (25 miles) across. The whole storm can be enormous. Hurricane Gilbert of 1988 was 1500 km (930 miles) across.

LIFE on EARTH

Life Begins

The most amazing thing that ever happened on earth took place about 4 000 000 000 years ago. Nothing like it has happened anywhere else (that we know of, anyway) in the whole universe. Life began.

We don't know how it happened, though we have a lot of ideas. If you had been able to visit the earth then, you might have been amazed that it happened at all. The air had no oxygen. There was no ozone layer. No animal could live there. But the first living things survived — and, in a few very strange places, something like them may still live today.

◀ WHERE DID LIFE BEGIN?
The first life needed energy. It also needed protection from the harmful rays of the sun. It could find both at the bottom of the sea.

In a few places in the sea, where sunlight cannot reach, boiling hot water pours out from deep in the earth. These volcanic vents are the only places on earth where life survives without the sun's energy. Tiny bacteria make food, using the vents' heat, and animals like the giant vent worms eat the bacteria.

Some scientists think that life may have started around vents like these. There was no ozone layer when life began, and the sun's ultraviolet rays would have destroyed anything trying to use sunlight for energy. The heat energy from the vents was safe.

VENT WORMS (*RIFTIA PACHYPTILA*), DEEP VOLCANIC VENTS IN THE PACIFIC

WHAT ABOUT...?
Why do we need the ozone layer?
See page 16.

PREVIOUS PAGES:
GIRAFFES (*GIRAFFA CAMELOPARDALIS*) AND AFRICAN ELEPHANTS (*LOXODONTA AFRICANA*), AFRICA

■ WHERE DID LIFE COME FROM?

Long chains of atoms called organic molecules are the building blocks of life. They had to form first, before life could begin.

The soot from a burning candle is made of carbon. Carbon atoms can join together to make long chains, like beads on a necklace. Other kinds of atoms, especially hydrogen, oxygen and nitrogen, can attach to them. Scientists think that the ocean was once full of these chains. Under the right conditions, they can grow by picking up more atoms, and perhaps can even make copies of themselves.

We know that these chains could form by themselves in the early ocean. But we have also found them in meteorites. Earth's first chains might have come from outer space — a sort of "space litter!"

AGGREGATE ANEMONES (*ANTHOPLEURA ELEGANTISSIMA*), PACIFIC COAST OF NORTH AMERICA

■ WHAT WAS THE FIRST LIFE LIKE?

We don't know just how the organic chains first joined together to make living things. The first life forms were tiny and very simple, like today's bacteria.

Bacteria are much smaller and simpler in the way they are organized than the cells in our own bodies. Fossils that look something like them have been found in 3 500 000 000-year-old rocks in Australia.

Some simple cells have been discovered living in water coming from volcanic vents. This water is hotter than boiling temperature, but cannot boil because of the weight of the ocean above it. Possibly they are leftovers from the earliest days of life on earth.

CORAL REEF, AUSTRALIA

HOW CAN I STUDY MICROSCOPIC CREATURES?

If you have a microscope at home or at school, you can easily watch live parameciums and other single-celled creatures. Almost any roadside ditch or stagnant pond contains millions of them. All you need is a jar to scoop up some of the water and pond scum, and an eyedropper to put a drop of the water on a microscope slide. The best kind of slide to use is called a well slide. It has a little bowl cut into the glass of the slide to hold the water. Float a cover slip over the drop, adjust the microscope's light, and you can begin your hunt.

Living Factories

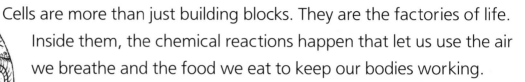

Your body is made up of millions of tiny building blocks called cells. But while real blocks are solid, cells are mostly liquid. They are something like plastic bags full of jelly. The "bag" is a covering called a membrane, and the "jelly" contains more membranes and a lot of complicated chemicals.

Cells are more than just building blocks. They are the factories of life. Inside them, the chemical reactions happen that let us use the air we breathe and the food we eat to keep our bodies working.

MICROSCOPIC CREATURES IN A DROP OF POND WATER

◀ **WHAT ARE CELLS LIKE INSIDE?**
Our bodies are made of separate parts or organs, like the heart and stomach. Plant and animal cells have separate parts, too. They are called organelles.

This paramecium is a microscopic living creature made of only one cell. It lives in freshwater ponds and ditches. A paramecium may look simple, but it is full of different organelles. Two of them, its nuclei, contain DNA, a chemical that "tells" the cell how to put itself together. It also has mitochondria to get energy from oxygen, lysosomes to digest its food, and many other types of organelles. On the outside of its body are hundreds of tiny hairs called cilia that the paramecium uses to row itself through the water.

Our cells have organelles, too. But they have one nucleus, not two, and because they don't swim by themselves, they are not covered with cilia.

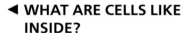

PARAMECIUM, WORLDWIDE IN FRESHWATER PONDS

■ **ARE ALL CELLS ALIKE?**
There are many different kinds of cells. For example, plant cells have hard walls around them, but animal cells do not.

You may think the most important difference between living things is the one between plants and animals. Plant cells have a hard wall of cellulose — the stuff that makes celery crunchy. They are missing some of the organelles animals have, and they have a few we don't.

But scientists think the biggest difference among living things is between cells without organelles (the prokaryotes) and cells with organelles (the eukaryotes). Prokaryotes were the first living things to appear, and for hundreds of millions of years they were the only life on earth. Bacteria are prokaryotes. Almost all other living things are eukaryotes.

WHAT ABOUT...?
Where did chlorophyll come from?
See page 51.

■ **WHERE DID OUR CELLS GET ALL THEIR PARTS?**
Some parts of cells like ours may once have been separate creatures. Thousands of millions of years ago, they joined together.

Imagine that hearts and stomachs once lived by themselves, then moved into animal bodies millions of years ago. It sounds weird, but that seems to be how cells like ours — the eukaryotic cells — got some of their organelles.

For example, cells do not grow their own mitochondria, the "lungs" of a cell. Instead, new ones are made when the old ones divide in half — the same way cells divide to make new cells. Mitochondria even have their own genes. In other words, mitochondria act like separate creatures that just happen to live in our cells. That makes us the result of a partnership that is thousands of millions of years old.

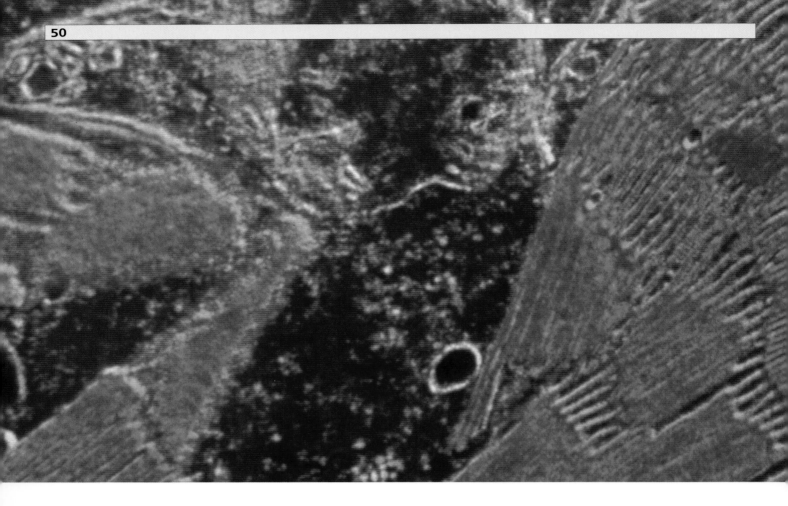

Why Green Plants Are Green

Many of the colors we see in animals or plants are there because of chemicals called pigments. Some of these chemicals do more than just make colors. For example, the pigment that makes carrots orange and flamingos pink is also a vitamin.

No pigment is more important to life on earth than the chemical that makes green plants green. It is called chlorophyll, and without it there would be no forests, no grasslands and no coral reefs. In fact, there would be no animals at all, and perhaps no living thing larger than a single cell.

WHAT ABOUT...?
**Why is there oxygen
in our air?**
See page 15.

KOALA (*PHASCOLARCTOS CINEREUS*) EATING
EUCALYPTUS (*EUCALYPTUS* SP.), AUSTRALIA

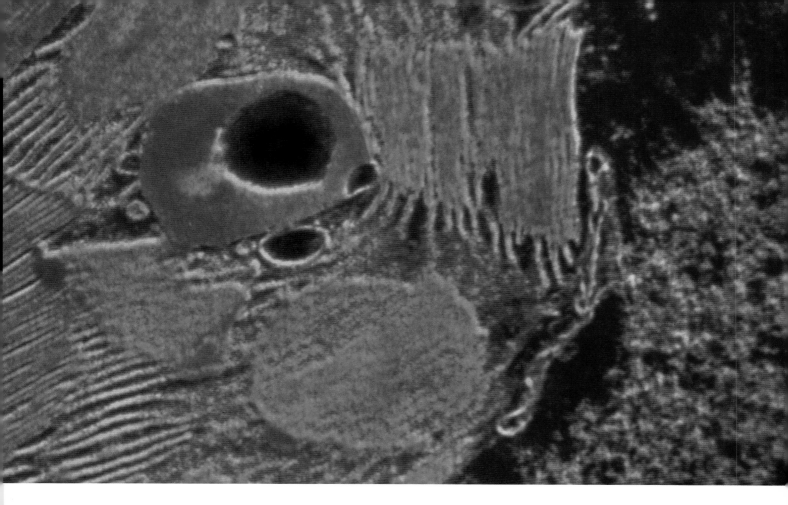

■ WHAT DOES CHLOROPHYLL DO?

Chlorophyll helps plants use the sun's energy to make food.

Water is made of hydrogen and oxygen atoms. Carbon dioxide is made of carbon and oxygen atoms. Green plants can use water and carbon dioxide to make sugars, which have carbon, hydrogen and oxygen atoms. But this takes energy, and the energy has to come from somewhere. Green plants get it from the sun.

Green plants need chlorophyll because it can absorb the sun's energy. Through a complicated set of chemical reactions called photosynthesis, the plant uses the energy the chlorophyll "collects" to build sugars. The sugars store the energy until the plant is ready to use it.

ABOVE: CHLOROPLAST, PHOTOGRAPHED WITH AN ELECTRON MICROSCOPE

▲ WHERE DID CHLOROPHYLL COME FROM?

The green color in plants is packed into tiny bodies called chloroplasts. Chloroplasts may have once lived on their own, but joined up with plant cells hundreds of millions of years ago.

The first living things to develop chlorophyll were not plants, but tiny cells called cyanobacteria. Some chloroplasts look a little like some cyanobacteria. Probably they once were cyanobacteria that joined up with larger cells to form partnerships. Today the chloroplasts use their chlorophyll to make food for the rest of the cell.

Partnerships like this one can also happen in the animal kingdom. Coral polyps, the animals that build reefs, have tiny green algae living in their bodies. The algae make most of the polyps' food, and the polyps provide a safe home for the algae.

■ WHY DO ANIMALS NEED PLANTS?

Animals can't get food energy from the sun. They can't put oxygen into the air. Green plants can. Without them, animals couldn't eat or breathe.

Photosynthesis stores the sun's energy in a form that plants can use, as the chemical energy that holds sugar molecules together. Animals can't do that, but they can get the energy another way. They can eat the plants. Once they have stored the energy from the plants in their bodies, other animals can get it by eating them. Without photosynthesis, both the plant-eaters and the meat-eaters would starve.

Not only would they starve, they would also suffocate. Photosynthesis gives off oxygen that is left over when the sugars are made. Without it, there would be no oxygen in the air for us to breathe.

The Book of Instructions

Some of the most wonderful parts of our bodies are far too small for us to see. The most amazing thing of all is a tiny molecule hidden deep in the nuclei of our cells. It is called deoxyribonucleic acid, but everybody uses its initials: DNA. DNA carries the genes — the instructions that tell the bodies of every living thing on earth how to grow, that determine what sort of creatures they will be, and that makes each of them different.

■ WHAT ARE GENES?

A gene is a plan your body can follow to do one thing, like color your eyes or skin, or grow straight or curly hair. Every living thing has thousands of different genes.

Imagine that you have to write the plans for a machine on a set of beads. The beads are so small that you can fit the plans for only one part of the machine on each one. But when you string the beads together, the string will carry the plans for the whole machine. That string is something like a DNA molecule. Each bead is like a single gene, and the machine is a living body.

Genes carry the instructions for making proteins. Proteins are the working parts that build and operate the body's machinery. Different proteins may help you digest your food, grow muscles or color your eyes.

WHAT ABOUT...?

What are mutants?

See page 59.

▼ HOW DOES DNA WORK?

DNA sends its instructions out into the cell by a special messenger called RNA. The RNA helps the cell make the proteins it needs.

The instructions on a DNA molecule are in code. There are four kinds of "teeth" along the DNA zipper: adenine, thymine, guanine and cytosine — or A, T, G and C for short. The order of these "teeth" along each side of the zipper spells out the code. It is like having an alphabet with only four letters. You could still

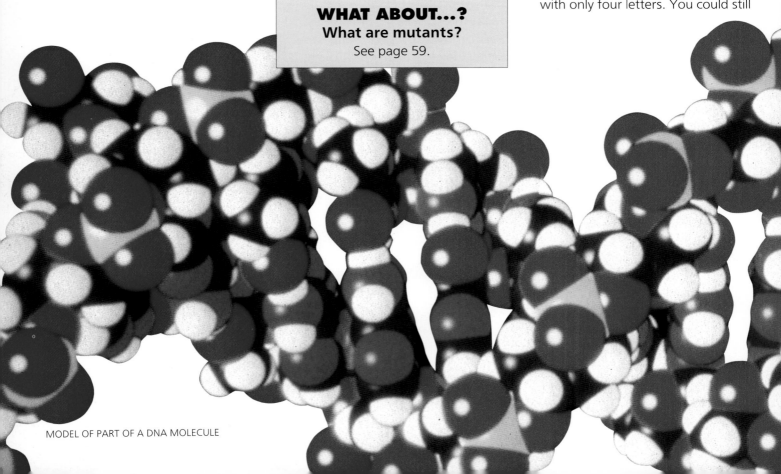

MODEL OF PART OF A DNA MOLECULE

spell out countless messages — ACGATAGCT, for example. Each message on a DNA molecule is a coded instruction for building one kind of protein. In other words, each DNA message is a gene.

DNA stays in the nucleus, but another molecule, RNA, "reads" the messages and carries them to the cell's protein-making factories. Here RNA acts like a pattern, lining up the bits that will make the protein in the right order.

■ HOW DO CHILDREN GET COPIES OF THEIR PARENTS' GENES?

A DNA molecule can split itself in two, and each part can make a copy of the other half. That is how parents' genes can be copied in their children.
DNA is shaped like a long, thin ladder twisted into a spiral. Each "rung" of the ladder is made of two pieces that lock together like the teeth on a

zipper. When a cell divides into two new cells, its DNA "unzips" itself. Instead of joining up again, each side uses chemicals in the cell to make a new half-zipper that fits it exactly. The result is two identical DNA molecules instead of one.

No other molecule can do that. Because DNA can copy its instructions, parents can pass copies of their own genes along to their children. Half of the genes in your body come from your mother, and half from your father. That is why you may look like your parents.

▶ WHY DON'T WE LOOK MORE LIKE CHIMPANZEES?

Chimpanzees and people have almost the same genes. But even a few different genes can make a big difference in what our bodies are like.
We may share more than 98 percent of our DNA with our closest relative, the chimpanzee. That's

more than chimpanzees share with gorillas! It may be, though, that the genes that are different are not the ones that code for specific body parts, but ones that make us grow and develop the way we do. These regulatory genes affect our whole body, and when they change they can make a big difference in the way we look. Even tiny changes can make a human baby grow into, for example, a very small or very large adult.

So the small difference in our genes makes humans and chimpanzees look different.

ARE CHIMPANZEES PEOPLE?

Some scientists think that chimpanzees and other apes are so much like us that they should be given the same rights that people have. That might mean that you couldn't keep them in zoos or do medical experiments on them. After all, would you put a person in a zoo? If you agree with this idea, you can help by writing about it to The Great Ape Project, P.O. Box 1023, Collingwood, Melbourne, Victoria, Australia 3066.

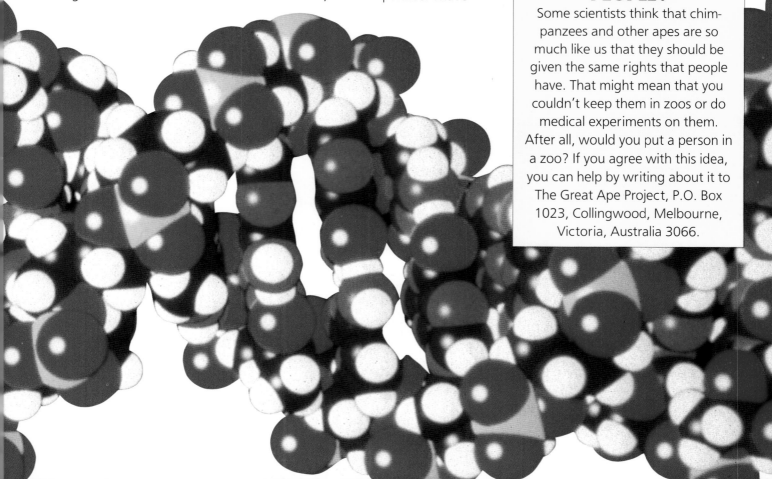

The Birds and the Bees

All living things can reproduce. In other words, they can make others of their own kind. Some split off bits of themselves that grow into new individuals. But most animals and plants reproduce by mating. When a male and a female mate, they produce young that have a mixture of genes, some from each parent. This is called sexual reproduction. Sexual reproduction is so important that many animals and plants spend much of their time and energy on it.

HONEYBEE (*APIS MELLIFERA*), ORIGINALLY AFRICA, EUROPE AND WESTERN ASIA

BEADLET ANEMONE (*ACTINIA EQUINA*) EXPELLING YOUNG, EUROPEAN SEAS

WHAT ABOUT...?
How do children get copies of their parents' genes?
See page 53.

▲ DO ALL ANIMALS OR PLANTS NEED TO MATE TO HAVE YOUNG?
Strawberries, sea anemones and many other animals and plants can grow young without mating. That helps them spread over a wide area.
Have you ever planted strawberries? Plant one and you will soon have a whole patch. These new plants don't grow from seeds. Instead, they bud from long stems called runners, sent out by the parent plant. They all have exactly the same genes. In fact, as long as they remain joined together they are really parts of the same plant.

Many plants, and some animals like this sea anemone, can reproduce by budding. It is a good way to produce many new individuals in a hurry without waiting for a mate to show up.

Bees produce young without mating, too. A hive needs lots of workers. Every worker honeybee hatches from an egg that was not fertilized by a male. Only the queens and drones hatch from fertilized eggs.

STILT-LEGGED FLIES (FAMILY MICROPEZIDAE) MATING, BAYER RIVER, PAPUA NEW GUINEA

▲ WHY DO ANIMALS OR PLANTS MATE?

When two animals or plants mate, their young will be different from their parents. They may be better at surviving in new places than their parents were.

Scientists have suggested a number of reasons why animals and plants go to the trouble of reproducing by sex. Young who are not identical to their parents may not catch the same diseases as easily. They also may be better able to move into new kinds of habitats, or survive better if their habitat changes. The truth is that we don't really know the answer to this question. All we know is that so many animals and plants use sexual reproduction that it must be very important.

▶ HOW DO PLANTS MATE?

Plants have male and female organs. The male organs make yellow, powdery pollen. Each pollen grain can fertilize a female "egg" or ovule.

Did you know that there are male and female pine cones? Male cones make pollen. Female cones make ovules, two on each cone scale. The male cones release their pollen into the air. The wind blows some of them to a female cone, where a few may land on a drop of sticky liquid on the ovule. Eventually one pollen grain grows down into the ovule and fertilizes it. This can take fifteen months!

The fertilized ovule becomes a seed. Half of its genes come from the pollen grain, half from the ovule. It is ready to grow into a new plant.

CONES FROM THE TAMARACK (*LARIX LARICINA*), NORTHERN NORTH AMERICA

What Are Flowers For?

Flowers and fruit are a plant's answer to its biggest problem: how to mate without moving. Pollen has to get from one plant to another, and once the seeds ripen, they have to move away from their parent. The wind carries the pollen and seeds of many flowering plants, as well as those of non-flowering plants like ferns and pine trees. Flowers and fruit, though, provide another solution: get an animal to carry the pollen or seeds instead.

RED GRANADILLA (*PASSIFLORA COCCINEA*),
NORTHERN SOUTH AMERICA

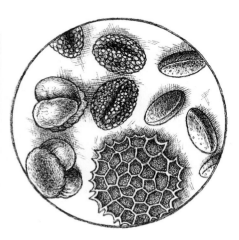

POLLEN GRAINS SEEN THROUGH A MICROSCOPE

◄ **WHAT IS A FLOWER?**
A flower is the part of a
flowering plant that makes its
seeds or pollen.

In the center of this granadilla flower are female carpels that make seeds, and male stamens that make pollen to fertilize the seeds. Surrounding them are brightly colored petals and sepals that attract butterflies. The butterflies carry the pollen to other flowers.

Not all flowers are like the granadilla. Some have no sepals or petals. Others have either male or female parts, but not both. Some are large and brightly colored, while others are tiny and dull. However, each flower design is adapted to spread its pollen in the way that suits its plant best.

"TOWHEAD BABIES" (FRUITS) OF WESTERN PASQUEFLOWER (*ANEMONE OCCIDENTALIS*), WESTERN NORTH AMERICA

WHAT ABOUT...?
How do plants mate?
See page 55.

◄ WHAT IS FRUIT FOR?
Fruit protects seeds and helps them to spread to new homes.

A fruit is an extra coating flowering plants grow around their seeds. It protects the seed from damage.

In many plants, it also helps the seed travel. Some fruits, like these "towhead babies," have parachutes that let them drift on the wind. Some float, like coconuts. Others are sticky or spiny, and can hitch a ride on a passing animal. Still others provide valuable food. Animals like us carry them away to eat, so their seeds fall far from the parent plant.

CONEFLOWER (*ECHINACEA* SP.) WITH HALICTID BEE (FAMILY HALICTIDAE), NORTH AMERICA

▼ WHEN IS A FLOWER NOT A FLOWER?
When it is many flowers, or a mixture of flowers and leaves, put together to look like a flower.

Some "flowers" are not what they seem. This coneflower is actually a mass of small flowers packed tightly together. The combination looks like a single flower.

The poinsettia "flower" popular at Christmastime is actually a group of small greenish flowers surrounded by bright red leaves. A combination like this does the same thing a single flower does: it attracts animals who end up carrying the plant's pollen.

Little Differences

Unless you are an identical twin, you don't look exactly like anyone else on earth. But you probably do look more like your close relatives than other people. You are more like them in other ways, too — ways that you can't see. For example, your blood type may be the same as theirs. Your children will share some of these special features with you.

Animals and plants differ from others of their own kind, too, though it may be harder for us to see differences in them than in people. These differences, like yours, are passed on from generation to generation.

MULE (*EQUUS CABALLUS* X *ASINUS*)

WHAT ABOUT...?
How does DNA work?
See page 52.

ORANGE AND WHITE TIGERS (*PANTHERA TIGRIS*), SOUTHERN AND EASTERN ASIA

▼ **WHY ARE WHITE TIGERS WHITE?**
White tigers are white because they carry a changed, or mutated, gene. They must get copies of this gene from both their mother and their father, or they will be orange. You may have seen white tigers at the zoo. They are like other tigers, except for their color. Their brothers and sisters may be orange, like most tigers, or white. White tigers can mate with other tigers, but unless they mate with another white tiger, most of their babies will be orange. That happens because if the baby carries the gene for orange color, that gene will work even if the changed gene is there, too.

Almost all of the white tigers in zoos today are the descendants of one white cub named Mohan, which was caught in India in 1951.

■ **WHAT ARE MUTANTS?**
Sometimes genes can change all by themselves. This is called mutation. The animal or plant with the changed gene is a mutant.

DNA, the molecule that carries our genes, normally makes identical copies of itself. Sometimes, though, something happens during the copying, and the new DNA is not the same as the old one. Bits of it may be changed, or left out or stuck back in upside down or in the wrong place. These changes are called mutations, from the Latin word meaning "change."

Most mutations make no difference. However, some may be harmful to the mutant carrying them. For example, some diseases are caused by mutations. But sometimes a mutant may be better able to survive than its parents, and can pass on its mutation to its children. That is one of the ways that evolution produces new forms of life.

▲ **WHAT IS A SPECIES?**
Animals or plants that can breed freely with one another belong to the same species, no matter how different they look.

There are more than 400 breeds of domestic dog. They may look very different, but they can all mate with each other and have puppies. They belong to the same species, *Canis familiaris*. Wolves and coyotes look much more like each other than like many domestic dogs, but do not usually breed together. They belong to different species, *Canis lupus* and *Canis latrans*.

If members of different species do breed together, the result is a hybrid. For example, a hybrid between a lion and a tiger is a liger, and a hybrid between a male donkey and a female horse is a mule.

ABOVE: EXAMPLES OF TEN DIFFERENT BREEDS OF DOG (*CANIS FAMILIARIS*)

■ **HOW DO NEW SPECIES FORM?**
If two groups of animals or plants can't reach each other, there is no way for their genes to mix. They may become so different that they could not breed together if they tried.

Animals or plants can't breed together if they live in different places. Sometimes a population of animals can be divided in two groups — perhaps by a river, a glacier or a mountain range. Each group continues to evolve, but they can no longer share their changes with each other. Later, the populations may meet again. If they have, in the meantime, become different enough, they may no longer be able to breed together. They will have become separate species. This is probably the commonest way new species are formed.

Survival of the Fittest

Wild apples are small and sour. Growers have worked for hundreds of years to change them into the fruit you eat. First, they selected and planted seeds from the largest and sweetest apples. When the new trees grew, the growers selected the seeds of the biggest and tastiest of those apples, and planted them. Finally, only big and tasty apples were left.

AN ORCHARD APPLE AND A WILD CRAB APPLE

The same thing happens in nature. Only animals and plants that live long enough to reproduce can pass on their genes to the next generation. Nature selects the ones best able to survive.

A STRANGLER FIG SURROUNDS A PALM TREE

INSIDE A STRANGLER FIG (*FICUS* SP.), WORLDWIDE IN WARM REGIONS

■ **WHAT IS EVOLUTION?**
Animals or plants may change, generation after generation, into different forms. This change is called evolution.
In 1857, a scientist named Charles Darwin published a famous book called *The Origin of Species*. In it, he showed how living things could evolve into new forms. As the environment changes, so do the animals and plants in it. If they don't change, or change in the wrong way, they may become extinct.

Darwin couldn't understand how an animal could pass its successful features on to its children. Today, we know that genes carry the instructions for new features. Only features coded in the genes, like eye color, can be passed on.

▲ **DO PLANTS FIGHT EACH OTHER?**
Plants fight each other for light and space. The strangler fig even kills other trees to get a place in the sun.
Forest trees need to be tall enough to reach the sunlight. Most of them have to grow from the ground up. The strangler fig has found a way to cheat.

It starts life as a vine that wraps itself around another tree. It uses the tree to climb quickly to the top of the forest. Most vines do that, but the strangler fig doesn't stop there. Its stems surround the other tree and grow together, walling it in like a prisoner. Eventually the other tree dies and rots away, leaving a hollow space inside the new fig tree.

■ WHAT IS "SURVIVAL OF THE FITTEST"?

Charles Darwin thought that animals and plants were always fighting one another in a struggle to survive. The ones that made it, and lived to raise young, were the "fittest."
That didn't mean that they really fought each other. They were more like runners in a race, where the best runners win and the poorest lose.

Today, many scientists think that this sort of struggle happens only in tough times. For example, if there is plenty of food for everybody, animals don't have to fight over it. But if food is hard to find, the animals that can find it most easily will have a better chance of living and passing on their genes.

▼ WHY DO PEPPER MOTHS COME IN TWO COLORS?

White pepper moths can hide more easily from hungry birds than black ones can. But if tree trunks are dirty from soot, the black ones can hide better.
The pepper moth is a famous example of evolution. Birds catch the moths that are easiest to see. White pepper moths are hard to see when they sit on a tree trunk. But in England, about 100 years ago, soot from factory chimneys stained the trunks of many trees black. Before long there were far more black pepper moths than white ones. Today factories are cleaner, and there are fewer dirty tree trunks. There are more white moths than black ones again.

This is how a whole population of animals can change when the environment changes. The ones that survive best after the change may not be the ones that survived best before.

PEPPER MOTHS (*BISTON BETULARIA*), NORTH AMERICA, EUROPE AND ASIA, ON A DARK TREE TRUNK (LEFT) AND ON A LIGHT ONE (RIGHT)

WHAT ABOUT...?
Are extinct species failures?
See page 66.

Evolving Together

When animals and plants evolve, the ways they change may affect other species. For example, if rabbits evolve ways to run faster, foxes will have to become faster, too, or they won't be able to catch the rabbits. Foxes can always find something else to eat. But what if rabbits were their only food? Then they would have to evolve with the rabbits or they would starve.

That sort of thing really happens. Animals or plants can become so involved in each other's lives that they evolve together. They may even become partners that cannot survive without each other. This is called co-evolution.

FOX CHASING RABBIT

◄ **WHAT IS LICHEN?**
A lichen looks like one plant, but it is really a partnership between an alga and a fungus. The fungus can't live without the alga.

Lichens grow like fuzzy mats on rocks or tree trunks. Tangled in the threads of the fungus are tiny, single-celled green algae. The algae can make food from sunlight, but the fungus can't. Some of the food the algae make is passed on to the fungus.

The fungus probably protects the alga and keeps it supplied with water and minerals. Strangely enough, though, the fungus can't live without the alga, but the alga can do quite well without the fungus.

STAGHORN LICHEN (*LETHARIA VULPINA*), WESTERN NORTH AMERICA

▶ WHAT IS A PARASITE?
A parasite lives on, or in, the body of an animal or plant, and gets food from it.

This harvestman has some unwanted passengers. They are tiny parasitic mites that live on it, like fleas on a dog. Some parasites live inside other animals. Your dog has probably been "wormed" to get rid of parasites that live inside its body. Some parasites are fairly harmless, but others can cause serious diseases.

There are even plants, like mistletoes, that parasitize other plants. Instead of making all their own food, mistletoes steal some of it from the plants they grow on.

Many parasites can live on, or in, only one kind of animal or plant. If its host changes through evolution, the parasite may have to change, too, in order to survive.

WHAT ABOUT...?
Who domesticated
the horse?
See page 82.

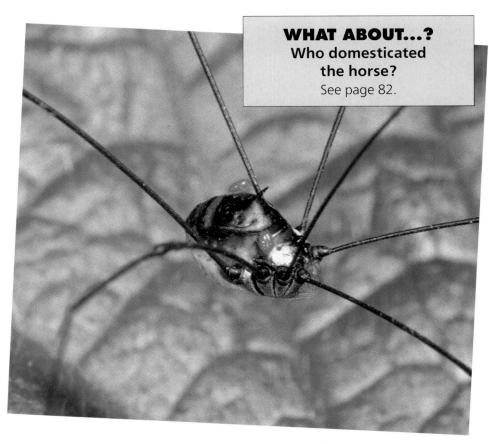

HARVESTMAN
(ORDER OPILIONES) WITH MITES
(ORDER ACARINA)

◀ WHAT PLANT GROWS A HOME FOR ANTS?
Ant plants like the bull's-horn acacia use part of their bodies as a home for ants. They even grow special ant food. In return, the ants protect the plant from its enemies.

Bull's-horn acacias grow in Africa and South America. They have thick, swollen thorns filled with spongy material. *Pseudomyrmex* ants burrow into the thorns, hollow them out and live there.

The acacia grows food for the ants on the tips of its leaves. In return, the ants act as the acacia's army. At the slightest touch they rush out to defend their home with vicious, painful stings. They kill any insect that tries to eat the acacia, and drive larger animals away. Brushing against an ant plant can be a very unpleasant experience!

BULL'S-HORN ACACIA (*ACACIA* SP.), AFRICA

Island Life

In 1835, the British ship *Beagle* sailed to the Galapagos Islands in the Pacific Ocean, on the equator, west of South America. On board was a young scientist named Charles Darwin. Darwin soon noticed that, although the animals and plants on the Galapagos looked like others from South America, they were actually different species. He then noticed something even stranger: not only were Galapagos animals different from South American ones, but each island had its own special kind of tortoise, iguana and other creatures.

WOODPECKER FINCH
(*CAMARHYNCHUS PALLIDUS*),
GALAPAGOS ISLANDS

GALAPAGOS TORTOISE (*GEOCHELONE ELEPHANTOPUS*),
GALAPAGOS ISLANDS

◄ **WHAT DID DARWIN LEARN FROM THE GALAPAGOS TORTOISE?**
Darwin noticed that each island had its own special kind of giant tortoise. He realized that they must all have evolved from one kind.
"Galapagos" is the Spanish word for "tortoise." The giant tortoises are the island's most famous animals. But it was not until Darwin's last day there that he noticed that tortoises from different islands had differently shaped shells.

Years later, Darwin realized what this meant. The tortoises once all looked alike, but they had changed over thousands of years to fit in with the special conditions on each island. The changed tortoises on each island couldn't breed with tortoises on the other islands. Each island's tortoises had changed in their own way. This discovery led Darwin to develop his theory of evolution.

WHAT ABOUT...?
How do volcanoes build the land?
See page 27.

MARINE IGUANA (*AMBLYRHYNCHUS CRISTATUS*), GALAPAGOS ISLANDS

BLUE-FOOTED BOOBY (*SULA NEBOUXII*), WORLDWIDE IN WARM SEAS

▲ WHY IS GALAPAGOS WILDLIFE SO SPECIAL?
Many Galapagos animals and plants are found nowhere else in the world.

Some animals are better at crossing the ocean than others. The ones that made it were able to take up new ways of life that, back in South America, were being lived by other animals that didn't make the crossing. There are no woodpeckers on the Galapagos. Instead, there is a woodpecker finch that uses a cactus spine to probe bark for insects. It is found nowhere else.

A few Galapagos animals have taken up very unusual ways of life. The marine iguana, for example, is the world's only ocean-going lizard. It feeds on seaweed.

■ ARE THERE OTHER PLACES LIKE THE GALAPAGOS?
Special animals and plants have evolved on islands around the world. Sadly, many of them are extinct, or in danger.

Hawaii, New Zealand, and the Mascarenes in the Indian Ocean are islands where animals and plants evolved strange new forms. For example, meat-eating caterpillars are found in Hawaii, and a parrot that can't fly in New Zealand.

Island animals had few enemies. When people arrived with dogs, cats, rats and pigs, many of these wonderful animals couldn't defend themselves and disappeared. The most famous of them all was a giant flightless pigeon from the Mascarenes. It became extinct so fast that we use its name today in a saying about extinction: "as dead as a dodo."

▲ HOW DID WILDLIFE REACH THE GALAPAGOS?
Animals and plants had to cross the ocean to reach the Galapagos. Some flew there, some drifted there and some were blown there.

When the volcanoes that formed the Galapagos Islands rose from the sea, they had no life. But soon after, tiny seeds and spores, carried on the winds, landed there and took root.

Later, animals arrived. Insects and birds may have been blown to the islands by storms. Lizards and tortoises may have floated from South America on logs or clumps of floating vegetation. Seabirds like these boobies flew to the islands. They may have brought passengers: insects hiding in their feathers, or seeds stuck to their feet.

How to Become Extinct

Almost all the animal or plant species that have lived on earth are now extinct. They died out because their environment changed. The change might have been big, like a forest turning into a desert, or small, like the arrival of a new kind of animal or plant that crowded the old ones out. Some species, though, evolved ways to survive changes in the environment.

SCYMNOGNATHUS, A MAMMAL-LIKE REPTILE FROM THE PERMIAN PERIOD

■ WHAT IS A MASS EXTINCTION?

When the dinosaurs disappeared, many other animals disappeared with them — almost half of earth's creatures in all. The loss of so many species in a short time is called a mass extinction.

Global changes have sometimes wiped out huge numbers of species. Before the dinosaurs died 65 million years ago, there were four other mass extinctions as bad or worse. "Smaller" ones have happened at least a dozen times. The last one was only 11 000 years ago, at the end of the Ice Age. Many species of large mammals, including the woolly mammoth, disappeared at that time.

Many scientists think that worldwide changes in climate, such as cooling or drying, caused the worst of the mass extinctions.

■ WHEN WAS THE WORST MASS EXTINCTION OF ALL TIME?

The worst mass extinction happened 245 million years ago — long before the dinosaurs. Nineteen out of every twenty species on earth were wiped out.

Near the end of what scientists call the Permian Period, the seas were full of life. On land, fierce predators like *Scymnognathus* hunted many different kinds of reptiles and amphibians.

But during this time, the world's continents were joining together into a single mass. When they did, the seas between them disappeared along with the creatures that lived in the water. Great deserts formed in the middle of Pangaea, the new supercontinent. The world's climate turned dry and cold.

Most of the species on earth vanished. The few survivors, though, kept evolving. A surviving relative of *Scymnognathus* gave rise, a few million years later, to the first mammals.

▶ ARE EXTINCT SPECIES FAILURES?

No. Many extinct animal and plant groups were successful for a long time. The dinosaurs lasted more than one hundred times longer than humans have done so far.

Many people think that, just because they finally died out, the dinosaurs were failures. But dinosaurs were, in fact, one of the most remarkable successes the world has ever seen.

For more than one hundred million years, they were one of the most important animal groups on earth. Dinosaurs lived on every continent, and they developed a huge range of shapes, sizes and ways of life. Remarkable animals like *Chasmosaurus*, one of the last of the dinosaurs, were still evolving close to the time they became extinct.

WHAT ABOUT...?
Are dinosaurs really extinct?
See page 73.

CHASMOSAURUS CANADENSIS

The Fossil Record

When people first noticed fossils, they had no idea what they were. The Chinese called them "dragon bones." In Europe, people thought that they were the remains of demons. Today, we know that fossils are the remains of animals and plants that lived millions of years ago. Without them, we would never have known that such wonderful creatures as dinosaurs had ever existed.

LIVING CRINOID (ORDER ISOCRINIDA), DEEP WATERS OFF THE BAHAMAS

WHAT ABOUT...?
What rocks have fossils in them?
See page 22.

MAMMOTH FOSSILS, HOT SPRINGS, SOUTH DAKOTA, USA

CRINOIDS, INDIANA, USA

▲ WHAT DO SCIENTISTS LOOK FOR WHEN THEY DIG UP FOSSILS?

Scientists do more than just dig up bones. They study the place where the bones were found.

These scientists are digging up a mammoth. There is a lot they would like to know that the bones can't tell them. Did the mammoth die there, or were its bones carried there by a rushing river? They look at how the bones are arranged, and they study the soil to see if it is the kind you would expect to find in a river. What kind of terrain did it live in? Scientists look for plant remains to see if the area was a forest, a grassland or a marsh. They may even look for signs of chewed-up food where the mammoth's stomach used to be.

■ HOW DO WE KNOW WHAT FOSSIL ANIMALS LOOKED LIKE?

We can study fossils of many animals and compare them with their living relatives. But, for some animals and plants, we just have to guess.

Most fossils show only part of an animal. Although we may never know exactly what the rest looked like, we can use other clues to give us a good idea. For a fossil like this crinoid, a relative of starfishes, it is easy because some crinoids are still alive and we can study them. For dinosaurs, we can study the muscles and skin of their relatives, the birds and the crocodiles, to help us decide what their muscles and skin were like, though we can never know what color they really were. Some fossils, though, are so different from anything living today that we can only guess what they looked like in life.

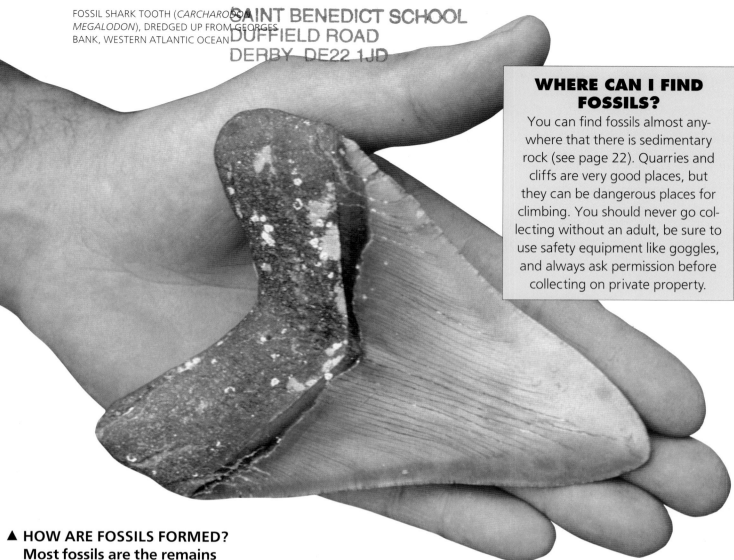

FOSSIL SHARK TOOTH (*CARCHARODON MEGALODON*), DREDGED UP FROM GEORGES BANK, WESTERN ATLANTIC OCEAN

(see page 22)

WHERE CAN I FIND FOSSILS?

You can find fossils almost anywhere that there is sedimentary rock (see page 22). Quarries and cliffs are very good places, but they can be dangerous places for climbing. You should never go collecting without an adult, be sure to use safety equipment like goggles, and always ask permission before collecting on private property.

▲ HOW ARE FOSSILS FORMED?
Most fossils are the remains of animals and plants that become covered by sand or mud. In time the sand turns to stone, sealing the fossil inside.

Usually, only remains that do not rot quickly last long enough to form fossils. That is why most fossils are of bones or shells, or teeth like this one. Shark teeth are common as fossils, but their skeletons, made of soft cartilage like the bridge of your nose, are very rare.

Many fossils not only are covered by stone, but also turn to stone themselves. Bit by bit, water carries dissolved minerals into the fossil, and washes fragments of the original bone or shell away. In time, minerals may replace the entire fossil with a perfect copy.

▶ WHAT DO PLANT FOSSILS TELL US?
Plant fossils tell us about changes in the environment over millions of years.

Plants don't have shells or bones. That's why plant fossils are harder to find than animal fossils. But plant fossils can be so well preserved that scientists can study their individual cells under a microscope. Plants, unlike animals, have hard cell walls that may not fall apart as the plant decays.

The commonest plant fossils are tiny grains of pollen. Studying them tells us whether the land where they grew was covered with trees, grass or other plants.

FOSSIL FERN

How to Live on Land

Life took more than three thousand million years to get out of the water. Water provides an ideal place to live. It keeps an animal or plant from drying out, buoys it up against the pull of gravity and protects it from extremes of heat and cold. But the land offered wonderful opportunities for plants and animals that could survive there: sunlight, food, and, at first, very few enemies.

The first plants and animals on land, like astronauts in space, couldn't survive without life support. They needed a way to keep their bodies moist. They needed to breathe air. They couldn't even hold themselves up, or move, without some sort of rigid skeleton.

JORDAN'S SALAMANDER
(*PLETHODON JORDANI*),
EASTERN USA

TOKAY GECKO
(*GEKKO GECKO*),
SOUTHEAST ASIA

■ WHAT DID ANIMALS NEED TO LIVE ON LAND?

Animals needed a way to stay out of the water full-time — even when very young. Reptiles solved the problem with a special kind of egg.

Most amphibians, such as frogs and salamanders, have to go back to the water to lay their eggs. However, the first reptiles evolved a new kind of egg, which is like a tiny spaceship. It is sealed against the outside world with a waterproof shell, and contains all the food and water the growing baby needs. Called an amniotic egg, it allowed reptiles, birds and early mammals to live far from the water. Today, though, some reptiles and almost all mammals give birth to living young instead.

▲ WHY ARE OUR LUNGS INSIDE OUR BODIES?

A land animal needs to keep its breathing apparatus moist. It usually does this by protecting it inside its body.

You would think that the best place for a breathing apparatus would be on the outside, where the oxygen is. But oxygen can get into our bodies only through thin membranes, which must be kept moist. That isn't a problem for animals with gills, like fishes. On land, though, gills can dry out.

Worms, salamanders and other animals solve the problem by staying in places where they can keep moist, and breathing through their thin skins. Reptiles like this gecko have thick, waterproof skins. Their lungs are inside their bodies, where they won't dry out as easily. So are ours.

▶ WHAT DID PLANTS NEED TO LIVE ON LAND?

Plants needed a way to grow upright and to carry food, water and minerals from one part of the plant to another.

Water plants can float upright, taking up dissolved minerals from the surrounding liquid. The first land plants had no roots or leaves, but they still had to reach down into the soil for water and minerals, and up toward the sunlight to make food. They needed something to support them, and to carry food, water and minerals back and forth.

Plants like this horsetail evolved stiff-walled tubes that did both jobs. They provided a skeleton and a transport system in one. In large plants, the tubes became stiffer still. Packed together, they formed the first wood.

FIELD HORSETAIL (*EQUISETUM ARVENSE*),
NORTH AMERICA, EUROPE AND ASIA

WHAT ABOUT...?
What do seashore animals do when the tide goes out?
See page 18.

Dinosaur News

Recent discoveries are changing many of our ideas about dinosaurs. For example, most dinosaurs did not drag their tails on the ground, as old books show them doing. Duckbilled dinosaurs probably took care of their babies, and at least one duckbill nested in colonies like a seabird. We are learning more about dinosaurs every year. But we still don't know why they disappeared — though scientists have lots of ideas.

▼ ARE NEW KINDS OF DINOSAURS STILL BEING DISCOVERED?
New dinosaurs are being found all the time. A new one is named about every seven weeks.

In 1984, scientists discovered a new dinosaur in Argentina. After years of study, they named it *Amargasaurus* in 1991.

Amargasaurus was a sauropod, a plant-eating dinosaur that lived about 125 million to 131 million years ago. It was smaller than most sauropods — about the size of an elephant. Unlike its relatives, it had a double row of long neck spines that may have supported a sail. *Amargasaurus* could have used its sail to show off to a mate or a rival. Blood flowing through the thin skin of the sail may been cooled by the air, helping to keep *Amargasaurus* from becoming too hot.

DINOSAUR FOOTPRINT

AMARGASAURUS CAZAUI, ARGENTINA

WHAT ABOUT...?
Did a meteor kill the dinosaurs?
See page 31.

► ARE DINOSAURS REALLY EXTINCT?
Many scientists think birds evolved from small, meat-eating dinosaurs. If they are right, then dinosaurs may not be extinct after all.

The earliest known birds had teeth and clawed hands. Their fossils look like those of small dinosaurs. Some bones of meat-eating dinosaurs are very bird-like. Some scientists even say birds are dinosaurs.

A large, flightless bird like this cassowary certainly looks like a dinosaur. A few scientists, however, think that birds are more closely related to crocodiles. They point to a newly discovered 225-million-year-old fossil called *Protoavis*. *Protoavis* is probably too old to be derived

NORTHERN CASSOWARY (*CASUARIUS UNAPPENDICULATUS*), NEW GUINEA

from a dinosaur, but it might be a bird. If it is — and many scientists think that it isn't — then birds may not be dinosaur descendants.

■ WHAT DO DINOSAUR FOOT-PRINTS TELL US?
Footprints can tell us where dinosaurs lived and how they stood and walked.

Millions of dinosaur tracks have been discovered around the world. Some scientists would rather study tracks than fossil bones, because tracks tell them how dinosaurs acted when they were alive. They know that most dinosaurs did not drag their tails on the ground, because there are no tail marks between the footprints. By measuring the distance between prints, they can estimate how fast a dinosaur was walking. Groups of tracks moving in the same direction tell them that many dinosaurs traveled in herds.

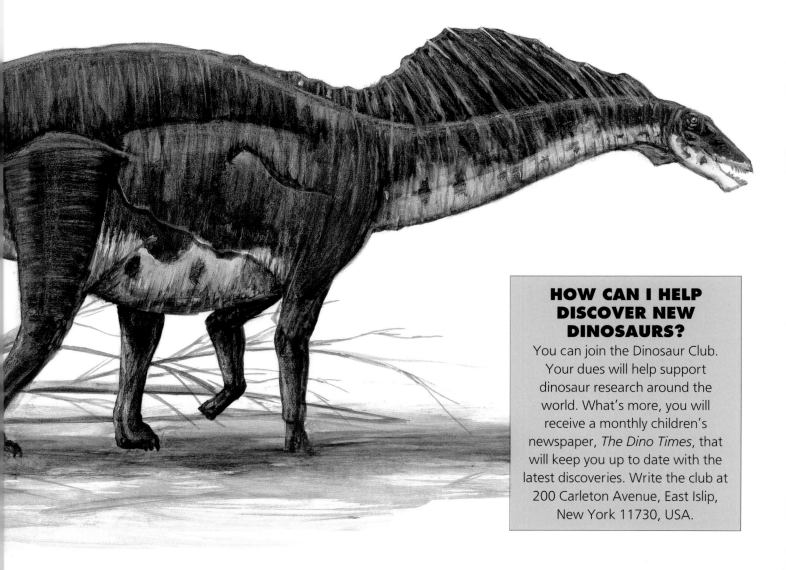

How to Be a Giant

Size makes a lot of difference to an animal. It affects how much food it needs, what its enemies are, how it moves, and where it lives.

Plants may become giant trees to reach the sunlight above a forest or to hold their leaves out of reach of hungry animals. Many giant animals, like the dinosaurs, are extinct. But the largest animals that ever lived, the great whales, are alive today. People are giants, too, compared with other animals. Most living things are much smaller than we are.

QUAKING ASPEN
(*POPULUS TREMULOIDES*),
NORTH AMERICA

■ **WHAT IS THE LARGEST LIVING THING?**
The largest living thing is a forest of 47 000 quaking aspen trees in Utah, USA. Because all the trees are joined together by their roots, they are really part of a single enormous plant.
Forests of quaking aspen are a common sight in the mountains of western North America. What you can't see as you walk through an aspen forest is that it is not a collection of separate trees. It is a single plant that has spread by sending out horizontal roots that may grow as much as 30 m (100 feet) before sprouting a new trunk.

The largest known aspen plant has been named "Pando," the Latin word for "I spread." It lives in the Wasatch Mountains of Utah. Pando has 47 000 trunks and may weigh more than 6 million kg (13 million pounds). It may be as much as a million years old, making it not only the largest living thing but perhaps the oldest as well.

HOW CAN I HELP PROTECT WHALES?
You can adopt one! If you join the International Wildlife Coalition's Whale Adoption Project, you will receive a picture of "your" humpback whale, an adoption certificate and a children's newsletter.
Your dues will help rescue stranded whales and support whale conservation. Write the International Wildlife Coalition, 634 North Falmouth Highway, Box 388, North Falmouth, Massachusetts, USA 02556.

■ WHY ARE THERE NO GIANT INSECTS?

Insect breathing systems only work for small animals. A giant insect wouldn't be able to breathe.

We use our blood to carry oxygen throughout our bodies. But insects have a system of tubes called tracheae running through their bodies. They pump air into the tracheae, where oxygen must diffuse slowly through the fluid in the tubes. This works well as long as the oxygen doesn't have far to go. It is much too slow to supply a giant — unless it has a thin body like the fossil dragonfly *Meganeura*. *Meganeura*, which lived 230 million years ago, was the largest known insect. Its wings were 70 cm (28 inches) across.

▼ HOW CAN WHALES GROW SO LARGE?

The biggest dinosaurs needed strong skeletons and muscles to hold themselves up, but water can support the weight of even the largest whale.

On land, very big animals have a problem — gravity. Even the largest dinosaurs known were much lighter than a 160-tonne (ton) blue whale. In the ocean, whales don't have that problem because water buoys them up. They still need vast amounts of food to supply their huge bodies, but their size helps them retain heat in icy waters.

A whale stranded on a beach, though, is in trouble even though it can breathe air. It may overheat in the sun. Because its skeleton is not designed to support its weight on land, the weight of its body squeezes the air from its lungs. A beached whale may die of exhaustion simply from trying to breathe.

WHAT ABOUT...?
Did people kill off the mammoths?
See page 82.

■ WHY ARE THERE NO REALLY TINY BIRDS?

Small, warm-blooded animals lose energy quickly. A really tiny bird would starve.

Birds and mammals are warm-blooded. They use food energy to produce heat. Because small objects cool off faster than large ones of the same shape, a small bird needs to burn more food energy to stay warm.

Hummingbirds, like the bee hummingbird, the smallest bird in the world, have to eat almost all the time. They risk starving to death in their sleep if they don't store enough food while awake. To prevent this, they can lower their body temperatures at night to save energy. A bird the size of a fly couldn't feed itself fast enough to stay alive.

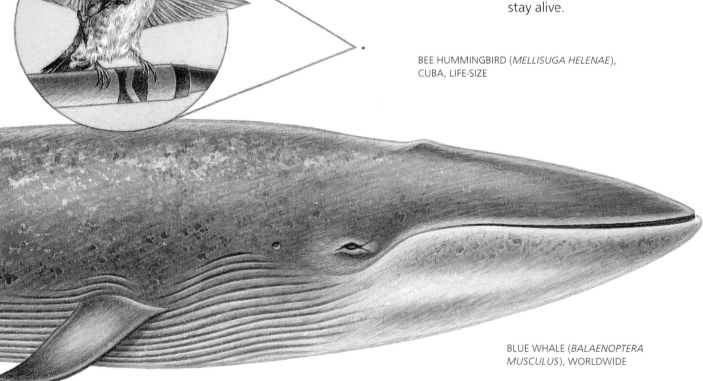

BEE HUMMINGBIRD (*MELLISUGA HELENAE*), CUBA, LIFE-SIZE

BLUE WHALE (*BALAENOPTERA MUSCULUS*), WORLDWIDE

World Records

When animals and plants set records, the prize they win is the only one that counts — survival. An animal or plant that can do something, or be something, better than any other living thing, has an advantage. Perhaps it can grow faster, or escape better, or have more young than its competitors.

But an animal or plant doesn't have to be a record-holder to survive. The second biggest animal isn't a loser. It's just better at being second biggest than the biggest animal!

ARCTIC TERN (*STERNA PARADISEA*), BREEDS ARCTIC, WINTERS ANTARCTICA

▲ WHAT BIRD MAKES THE LONGEST JOURNEY?
Every year the arctic tern flies from the Arctic to the Antarctic and back again — a distance of 17 600 km (11 000 miles) each way.
One arctic tern, marked with a metal band on its leg in Russia, was caught again less than a year later in Australia, 22 400 km (14 000 miles) away.

Summer days in the Arctic and Antarctic are very long. Because the arctic tern spends so much time in both the arctic summer and the antarctic summer, it may see more daylight per year than any other animal.

BRISTLECONE PINE (*PINUS ARISTATA*), WESTERN UNITED STATES

▲ WHAT IS THE OLDEST LIVING TREE?
The oldest living single tree is a bristlecone pine.
The oldest living thing may be a clump of quaking aspens (see page 74). But the oldest living single tree is a bristlecone pine in the White Mountains of California. It is at least 4600 years old.

Like most trees, bristlecone pines produce a new ring of wood each year. Scientists can take a thin core of wood from a tree, count the rings, and tell exactly how old a tree is. Rings from bristlecone pines have been used to check the age of human artifacts. Scientists compare the amount of radioactive carbon found in the rings and in old objects. Such studies have proven that ancient monuments, such as Stonehenge in England, are centuries older than we had thought they were.

▶ WHAT IS THE WORLD'S FASTEST RUNNER?

A cheetah can run at 110 km/h (68 mph), but it can't keep it up for very long.

The cheetah, a member of the cat family, is built for speed but not for endurance. If it doesn't catch its prey within 300 m (1000 feet), it has to give up. When cheetahs hunt, they try to get as close as possible to their prey before they charge — usually within 50 m (150 feet) or less. They may take as much as an hour trying to get into a good position. Even then, they may only catch their victims a little more than half the time.

CHEETAH (*ACINONYX JUBATUS*), AFRICA AND SOUTHWESTERN ASIA

■ WHAT ANIMAL LAYS THE MOST EGGS?

The ocean sunfish may lay millions of eggs at one time.

Ocean sunfishes are huge — they may weigh up to 1500 kg (3300 pounds) — but their eggs are each smaller than a pinhead. One 1.24-m (4½-foot) female was thought to have 300 million eggs inside her body.

■ WHAT IS THE WORLD'S BIGGEST SEED?

A Seychelles double coconut may weigh 20 kg (44 pounds).

The double coconut is the seed of the coco de mer, a palm tree from the Seychelles Islands in the Indian Ocean. Hundreds of years ago, travelers thought the seeds, found washed up on distant shores, were antidotes for any kind of poison. European kings and queens paid huge sums of money for them.

WHAT ABOUT...?
What is the oldest kind of tree in the world?
See page 79.

MAYFLY (ORDER EPHEMEROPTERA), WORLDWIDE

◀ DO MAYFLIES REALLY LIVE FOR JUST ONE DAY?

You may have heard that mayflies live for one day. But that isn't really true. Adults may live only a day, but young mayflies can live for years.

Once they hatch, mayflies start their lives as wingless larvae, or naiads, living at the bottoms of lakes and streams. Some naiads may live for four years. When it comes time to mate, they crawl out of the water, shed their skins twice, and emerge by the thousands as adults. The delicate adults can fly, but they can't eat. They have only a few hours, or at most a few days, to mate and lay eggs before they die.

Living Fossils

A living fossil is an animal or plant that still looks the way it did millions of years ago. Some living fossils, like the cockroach, are still found around the world. Others, like the ginkgo tree, now live only in a few special places. The most exciting living fossils are those that were once thought by scientists to have been extinct for millions of years — and then turned out to be alive after all.

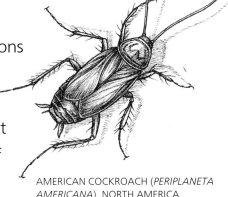

AMERICAN COCKROACH (*PERIPLANETA AMERICANA*), NORTH AMERICA

COELACANTH (*LATIMERIA CHALUMNAE*), COMORO ISLANDS

▲ WHAT IS A COELACANTH?
The coelacanth (seel'-a-kanth) is the last of the lobe-finned fishes. It may be more closely related to us than any other fish.

In 1938, fishermen caught a strange fish off the coast of South Africa. It was a coelacanth — a fish thought to have been extinct since the days of the dinosaurs. Excited scientists spent fourteen years trying to find another one. We now know that coelacanths live in deep water around the Comoro Islands in the Indian Ocean. The local people knew about them all along, of course. They had been catching them in their nets for centuries.

About 300-million years ago, a fresh-water cousin of the coelacanth crawled out onto the land, using its fleshy fins to push itself along. It became the ancestor of all land vertebrates — amphibians, reptiles, birds and mammals.

GINKGO (*GINKGO BILOBA*), CHINA, PLANTED WORLDWIDE

GREEN DARNER (ANAX JUNIUS), NORTH AMERICA

CHAMBERED NAUTILUS (*NAUTILUS* SP.), DEEP WATERS OF THE WESTERN PACIFIC

▲ WHAT IS THE OLDEST KIND OF TREE IN THE WORLD?

The oldest kind of tree is the ginkgo. It has no close relatives. The ginkgo tree has survived almost unchanged for 200 million years.

Ginkgo trees grow in many of our cities today. Their fan-shaped leaves are very pretty, and they can live in polluted air that kills other trees. But the gingko is no ordinary tree. It is a true living fossil.

Like pine trees, ginkgoes have no fruit or flowers. In the time of the dinosaurs, before flowers even existed, they were found all over the world. Today there is only one kind left. The last few wild ginkgoes live in a tiny corner of southeastern China. The Chinese planted them around their temples, and the trees in our own parks come from seeds collected in those temple gardens.

▲ IS THE DRAGONFLY A LIVING FOSSIL?

Dragonflies are living fossils that have changed very little in 200 million years.

Not all living fossils are rare. The very oldest dragonflies we know lived about 230 million years ago. Although they are older than the dinosaurs, there are about 5000 different kinds of dragonflies flying about the world today. They were one of the earliest animals to fly, and still have some features from those long-ago days.

WHAT ABOUT...?
What is "survival of the fittest"?
See page 61.

▲ WHAT DOES THE CHAMBERED NAUTILUS TELL US?

The chambered nautilus's relatives became extinct 65 million years ago. The living nautilus shows us what they probably looked like.

The chambered nautilus is a cousin of octopuses. Unlike them, it lives in a beautiful shell. As the nautilus grows, it moves forward into a new, larger section of the shell, and seals off the old behind it with a pearly wall. A nautilus shell is divided into many of these chambers. These chambers, which are mostly full of gas, help the nautilus hang in the water as it swims.

There were once hundreds of kinds of nautiluses and their relatives, the ammonites. We have only their shells as fossils. If the chambered nautilus had not survived, we might not have known what the creatures that made the fossil shells were like.

WHAT ABOUT...?
Why don't we look more like
chimpanzees?
See page 53.

**HOW CAN I HELP
SAVE PRIMATES?**
The Fort Wayne Children's Zoo is
trying to save five kinds of
primates that live only on the
Mentawai Islands in Indonesia. It
has helped 12 000 school children
raise money to set up a rainforest
reserve in the islands. Letters from
children to the Indonesian govern-
ment have helped persuade it to
stop cutting down trees where the
primates live. Your teacher or
group leader can get a free
resource packet by writing to the
Mentawai Islands Conservation
Project, c/o Fort Wayne Children's
Zoo, 3411 Sherman Boulevard,
Fort Wavne. Indiana 46808, USA.

Where We Come In

Around the time of the last of the dinosaurs, there lived in the Rocky Mountains of North America a little animal called *Purgatorius*. We don't know what *Purgatorius* looked like, because all we have are its fossilized teeth. Those teeth, though, are enough to show that *Purgatorius* may have been a primate — the oldest one known. And that makes *Purgatorius* very interesting, because we are primates, too. It took almost 65 million years for some African relatives of *Purgatorius* to evolve into the first human beings. We humans are not the only primates alive today. Lemurs, monkeys and apes are our cousins.

RING-TAILED LEMUR (*LEMUR CATTA*), SOUTHERN MADAGASCAR

GOLDEN LION TAMARIN (*LEONTOPITHECUS ROSALIA*), SOUTHEASTERN BRAZIL

◄ WHAT ARE OUR CLOSEST LIVING RELATIVES?

Our closest living relatives are the great apes: the chimpanzee, the bonobo or pygmy chimpanzee, the gorilla and the orangutan.

Apes are not monkeys. The great apes, and their cousins the gibbons, use their arms to swing about in the trees. Monkeys don't get around that way. They leap with all four feet, although a few South American monkeys can swing by their tails.

The great apes are much more like us than monkeys are. Like people, apes do not have tails. They have large brains, and are very intelligent. We are, in fact, a kind of great ape ourselves. Some scientists think that we are really a sort of chimpanzee.

GORILLA (*GORILLA GORILLA*), CENTRAL AFRICA

▲ WHERE IS THE ISLAND OF GHOSTS?

Madagascar is the home of the lemurs. Their large yellow eyes make these animals look like ghosts to some people. "Lemur" is Latin for "ghost."

Before the first monkeys evolved, their older cousins the lemurs lived almost around the world. Today, the only place true lemurs still live is on the giant island of Madagascar, off southeastern Africa, and a few smaller islands nearby.

Monkeys are smarter and more aggressive than lemurs, and probably drove them from the rest of the world. But monkeys never reached Madagascar. Today, twenty-three species of lemur live there. We don't know how long they will last. The forests of Madagascar, the lemurs' only home, are among the most endangered on earth.

▲ WHY ARE SO MANY PRIMATES IN DANGER?

Many primates live in tropical rainforests. People are destroying rainforests around the world.

More than fifty species of primates are in danger of extinction. Many others are growing rare, mostly because their homes are being destroyed. Hunters still trap rare monkeys and apes to sell as pets or for research, though this is now illegal in most countries.

The golden lion tamarin lives in the Atlantic forest of southeastern Brazil. Almost all of this forest has been cut down, and the primates that live there have become very rare. People are trying to save golden lion tamarins by protecting what is left of their forest home, and by raising more of them in zoos.

People Make a Difference

Human beings evolved in Africa, but we didn't stay there. Wherever our ancestors went they changed the lives of other animals. Many spectacular creatures became extinct — possibly killed off by early man. A few were changed from being our prey to being our pets or our work animals. With the invention of agriculture, we began to change the land itself. As people took over more and more of the earth, the habitats of wild animals began to disappear.

GIANT MOA (*DINORNIS GIGANTEUS*), NEW ZEALAND (EXTINCT)

PRZEWALSKI'S HORSE (*EQUUS PRZEWALSKI*), CENTRAL ASIA

▲ WHO DOMESTICATED THE HORSE?
Horses were probably first tamed over 5000 years ago, in central Asia.

Early humans hunted horses for food. They painted their pictures on cave walls 15 000 years ago. Probably the first people to tame horses were the Scythians of southern Russia, around 3100 BC. By 4000 years ago, domestic horses were common in China.

Their wild cousins, though, have not done so well. The last kind of truly wild horse, the Przewalski's horse of central Asia, survives only in captivity. Scientists are now planning to reintroduce Przewalski's horses into their former home. All other "wild" horses are descended from domestic horses that have escaped from humans and returned to the wilderness.

▶ DID PEOPLE KILL OFF THE MAMMOTHS?
Early humans probably hunted mammoths, but they may not be to blame for their extinction.

The woolly mammoth, the hairy elephant of the Ice Age, is only one of many large animals that disappeared after the arrival of humans. It became extinct almost everywhere about 10 000 years ago, though dwarf mammoths, a third the weight of the normal ones, may have survived on Wrangel Island off northern Siberia for another 6000 years.

As the great ice sheets melted and the climate changed, mammoths were probably already growing rare. Some scientists think that human hunters may have killed the last of them. Others think humans rarely hunted mammoths. We may never know for sure. We do know, though, that humans did kill off some giants, such as the moas of New Zealand — birds up to 3 m (9 feet) tall.

WOOLLY MAMMOTHS (*MAMMUTHUS PRIMIGENIUS*), NORTH AMERICA AND EURASIA (EXTINCT)

WHAT ABOUT...?
Why are so many animals
in danger?
See page 90.

HELPING the EARTH

HOW CAN I HELP FIGHT POLLUTION?

There are many ways to fight pollution. The reading list at the back of this book includes books that will help you. You can start with simple things — like turning off lights or the TV when you leave a room to save energy, not leaving taps on for a long time to save water, and following the three R's. If you want more ideas or information, you can write or telephone your local environmental group.

WE
RECYCLE

WHAT ABOUT...?
How can I help save wetlands?
See page 37.

Fighting Pollution

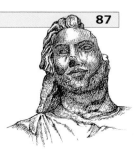

STATUE DAMAGED
BY ACID RAIN

Today, many of us live more comfortable lives than the kings and queens of olden times. We have luxuries they never dreamt of — and we are paying a price for it. That price is a polluted world.

But we don't have to pollute. We can stop poisoning the land with pesticides and chemical fertilizers. When we are finished with our possessions, we can recycle them and the packages they came in, instead of throwing them into higher and higher mountains of garbage.

A FARMER SPRAYING PESTICIDES

▲ WHAT CAN WE DO ABOUT GARBAGE?

Each Canadian throws away about a tonne of garbage per year —the weight of a small car. We should all buy less, use things over again, and recycle.

We are running out of places to put all our garbage. We are taking good land that could be used for farms or parks and turning it into garbage dumps. Many large cities have used up all the landfill sites — places to dump garbage — nearby, and have to truck their trash a long way. A lot of our garbage contains things we never needed in the first place. One-third of it is nothing but packaging!

Garbage is hard to get rid of. Plastic garbage will last for centuries. Some can be burned, but that can give off dangerous gases. The best way to solve our garbage problem is to follow the three R's: Reduce, Reuse and Recycle.

■ WHAT IS ACID RAIN?

Chemicals from car engines and factory chimneys combine with water vapor to make acid. When the water falls as rain, the acid in it can damage lakes, forests and even buildings.

Acid rain can "kill" a lake, because many animals and plants in lakes can't survive if the water becomes too acidic. It is also killing trees in Europe and North America. Since acid can dissolve limestone and marble, acid rain eats away at buildings and outdoor statues, too.

Catalytic converters in cars and sulphur "scrubbers" in factories can remove some harmful chemicals before they can cause acid rain. Governments are trying to find more ways to reduce this kind of pollution.

▲ WHAT IS BAD ABOUT PESTICIDES?

Pesticides don't just kill pests. They poison the air and water, and kill the tiny creatures that make the soil rich.

Pesticides are deadly poisons. They soak into the soil, killing earthworms and other valuable animals. Rain washes them into rivers and lakes, where they poison the water.

Pesticides are supposed to kill insects that eat crops, or mosquitoes that carry diseases like malaria. But insects evolve so quickly that hundreds of them have become immune to the poisons. The result is that, worldwide, we still lose as much of our crops to insects as we did before we started using pesticides.

Today many people try to grow food without using chemical pesticides.

A Change in the Air

We humans are changing our planet's air. Some of these changes may be very dangerous. There are holes in the ozone layer that protects us from the sun's ultraviolet rays, and gases we add to the air by burning forests and driving cars may be warming our climate.

The problems in our air may seem too big for anyone to handle. But although we probably can't stop global warming right away, or close the holes in the ozone layer today, we can make a start.

■ WHAT IS THE GREENHOUSE EFFECT?

The windows of a greenhouse trap the sun's heat inside. Greenhouse gases in our air do the same thing for the whole world.

Carbon dioxide, methane and water vapor are greenhouse gases. They soak up the sun's heat before it can bounce off the earth back into space. That warms up the air, and some of that warmth goes back to the ground. This is called the greenhouse effect.

If the earth had no greenhouse gases in its air, most of the heat that it gets from the sun would escape into outer space. The earth would be much colder than it is today: −19°C (2°F) on average instead of +14°C (57°F).

■ HOW DOES GLOBAL WARMING HAPPEN?

People are putting more and more greenhouse gases into the air. More greenhouse gases means that more of the sun's heat will be trapped.

Greenhouse gases like carbon dioxide make up only a tiny part of our air — less than one part in three hundred. People are adding more carbon dioxide to the air all the time, though, and that tiny part is getting bigger. Every time we burn something, from the gas in a car's engine to a tropical forest, we add carbon dioxide to the air.

■ WHY IS GLOBAL WARMING DANGEROUS?

If our climate gets much warmer, many of our crops may die. Thousands of animals and plants may become extinct.

Farmers need the right kind of climate to grow their crops. A good climate for wheat, for example, may not be right for rice. If our climate changes, many of the crops we need for food may not be able to grow where farmers have always grown them.

Farmers can move, or grow new crops, but wild animals and plants may be stuck where they are. Many will die if the climate they are used to changes.

WHAT ABOUT...?
Why doesn't the earth get too hot?
See page 16.

■ WHAT IS HAPPENING TO THE OZONE LAYER?

Chemicals called CFCs drift up high in the sky. There, they turn ozone into ordinary oxygen. The ozone layer is getting very thin or is disappearing in many places.

In 1981, scientists in Antarctica discovered a hole in the ozone layer over the South Pole. Over the next few years the hole grew bigger. Another hole was found over the North Pole in 1989. Today, the ozone layer is getting thin over many large cities, too.

The culprits are chemicals called chlorofluorocarbons, or CFCs. We use CFCs in refrigerators and air conditioners. Some countries may still be using them in spray cans. In 1990, many countries agreed to stop using CFCs by the year 2000. Once they do, the ozone holes should slowly repair themselves over the next century.

RUSH-HOUR TRAFFIC, TORONTO, ONTARIO, CANADA

CAN I DO ANYTHING TO HELP?

Remember that every time you ask your parents to drive you somewhere when you could walk or bicycle, you are adding greenhouse gases to our air. Get there under your own power instead. That's a good way to start saving our climate — and it's good exercise, too!

Using Up the World

Our planet is a wonderful place. It is full of millions of kinds of living things. Every one of them is special in some way. The forests, coral reefs and other places where they live are beautiful and precious. But we are using them up so fast that, unless we change our ways, many of our animals and plants, and the places where they live, will be gone by the time you grow up. It's up to us to make sure that doesn't happen.

ATLANTIC RAINFOREST BEING CLEARED, BAHIA, BRAZIL

PALM COCKATOO (*PROBUSCIGER ATERRIMUS*), NEW GUINEA AND NORTHERN AUSTRALIA

▲ WHY DO WE DESTROY FORESTS?

People cut down forests to sell the wood, or to make room for farms or ranches. When the forest is gone, the land that is left may be worth nothing.

Forests contain many treasures. Rainforest plants may contain cures for diseases like cancer. But instead of carefully looking for these riches, logging companies turn whole rainforests into products like decorative paneling for television sets or disposable forms for pouring concrete.

We are trying to learn to use nature in ways that do not destroy it. But some places, like old-growth forests and coral reefs, will always be too delicate to use. They must be protected for the sake of the animals and plants that live there and the people who come to enjoy their beauty.

■ WHY ARE THE DESERTS GROWING?

The deserts of the world are spreading because people are turning farmland into places where few plants can grow.

Wild places like rainforests are not the only parts of the world we are using up. Every year, we turn 27 000 000 ha (675 000 000 acres) of farmland — an area the size of the Australian state of Victoria — into desert. We do it by using bad farming methods, and by cutting down trees that should be protecting the soil. In Africa and Asia, millions of people are going hungry because they can no longer grow food.

In many places, villages and farm communities are fighting the growing deserts. The people are planting new trees, or even putting up rows of stones to keep the soil from washing away.

■ WHY ARE SO MANY ANIMALS AND PLANTS IN DANGER?

Some are in danger because we want to use them, and we take too many of them. Many more are in danger because people are destroying their homes.

Many people want beautiful pet parrots, flowering orchids, or jewelry made of elephant ivory or coral. Some people, particularly in Asia, use bear gall bladders and other animal parts as medicine. When hunters take too many animals or plants from the wild to sell to these people, the animals or plants may start to disappear.

But most endangered species today are dying because we are destroying the habitats where they live. When we cut down a rainforest or pollute a coral reef, thousands of animals and plants die.

OPPOSITE: AMAZON RAINFOREST, PARA, BRAZIL

WHAT ABOUT...?
How do rainforests cool the world?
See page 38.

HOW CAN I STOP USING UP THE WORLD?

You may not be old enough to buy many things, but your parents are. You can ask them not to buy products that use rainforest timber, like television sets with rosewood paneling. If you travel, don't buy souvenirs made from coral or seashells (but it's okay to pick up empty seashells on the beach!). Look at the reading list for books that will tell you how to become a caring shopper. And always remember — don't waste the things you do have. That paper you toss in the garbage was once part of a living tree.

How Can I Help?

Rescuing the earth will be a long, hard job, but it is a job everyone can do. You don't have to be a famous scientist or a world leader to help. You don't even have to be a grownup. When leaders from around the world met in Brazil in 1992 to try to find ways to save the world, one of the most important speeches was made by Severn Cullis-Suzuki, an 11-year-old girl.

You have already seen, in the earlier pages of this book, some ways you can help. You can probably think of others yourself, or find them in books or newspapers or on television.

FILLING CONTAINERS FOR TREE SEEDLINGS

■ CAN I REALLY HELP?
Children have helped save tropical rainforests, rescue injured animals, plant trees and clean up the environment.
Problems like global warming or the destruction of the rainforests may seem too big for anyone to solve. But even the biggest problems are caused by many people each doing some little thing that is bad for our earth. So if many people, including you, can each do some little thing that is good for the earth, then the problems will start to be solved. The more good things we can do, the better it will be. It doesn't matter where you start, or what you decide to do. The important thing is to believe that you **can** help.

■ HOW WILL I KNOW WHAT TO DO?
The most important thing you can do to help our planet is to learn all you can about it.
The more you know about what is happening to our earth, the more you will be able to help to save it. Books like this one are a good place to start. But if you want to find out what is going on right now, you should look for stories about wildlife in your newspaper, or on the radio or television news.

If you hear about something bad happening to our air and water, to wild animals or plants or to the places where they live, tell your parents or your teacher. Write a letter to your member of Parliament or Congress, telling him or her how you feel. Perhaps your whole class can write letters and send them together. You can make a real difference!

Glossary

Here are the meanings of some of the words in this book. Words in **bold face** *have their own entries in this glossary.*

Atmosphere: The blanket of air that surrounds the earth. Our atmosphere is 78 percent nitrogen and 21 percent oxygen. Other gases, such as argon and carbon dioxide, make up the remaining 1 percent.

Bacteria: Tiny living things, each made of a single cell that has no nucleus. They are found everywhere on earth, from the deepest seas to high in the atmosphere.

Desert: A place that receives less than 25 cm (10 inches) of rain or snow per year.

Endangered species: An animal or plant that is in danger of becoming **extinct** soon.

Erosion: The process by which wind, water, ice and gravity wear away rocks, break them into tiny pieces, and carry away pebbles and soil. Erosion can flatten mountains and dig canyons through level ground.

Evolution: Change in animals and plants that is passed on from generation to generation.

Extinct: A **species** of animal or plant becomes extinct when the last one dies.

Fault: If the rock on either side of a crack in the earth's crust has moved, the crack is called a fault.

Fossil: The remains of an animal or plant that lived in the distant past. A fossil can even be the remains of a footprint or a burrow; then it is called a trace fossil.

Genes: Genes are the instructions for building our bodies and making them work. One gene is the length of DNA needed to carry the instructions for a single piece of protein, called a polypeptide.

Gravity: A very weak force that attracts any object toward the center of another object. The more massive the object, the stronger its gravity. The earth is so big that its gravity can keep our air, our water and our bodies from flying off into space. The pull of the earth's gravity gives us weight.

Lava: Molten rock that comes out of a volcano. If it is still underground, it is called magma.

Mating: Animals mate when the sperm of the male joins with the egg of the female to produce a new individual.

Molecule: The smallest piece of a chemical that still acts like that chemical. It is made of several atoms bonded together. For example, one carbon dioxide molecule is made of one carbon atom and two oxygen atoms. It is still carbon dioxide, but if you take it apart, it is just carbon and oxygen.

Ozone layer: A layer in our atmosphere, about 15 to 30 km (9 to 19 miles) up, which contains large amounts of ozone gas. The ozone layer absorbs part of the ultraviolet radiation from the sun that strikes the earth.

Photosynthesis: A process used by green plants to make sugar out of water and carbon dioxide, using energy from the sun.

Planetesimal: One of the "space rocks" that collided with one another to form the planets and moons during the birth of our solar system.

Pollination: Pollination happens when the male pollen joins with the female ovule to form the beginning of a new seed.

Predator: An animal that kills and eats other animals.

Prey: An animal that is eaten by a **predator**.

Rainforest: A forest where there is a lot of rain all year round. Its trees stay green all year.

Species: A group of animals or plants that can mate with each other and have babies, but usually cannot mate with members of other species. Usually, when we talk about a "kind" of animal or plant, like a tiger or a sugar maple, we mean a species.

Scientists give species scientific names made up of two words. The first word is the genus, a collection of species that are related to one another. The second word is the species. For example, the lion and the tiger are different species in the same genus. The lion's scientific name is *Panthera leo* and the tiger's is *Panthera tigris*.

Tectonic plate: One of the large sections of our planet's crust. The movement of the plates causes the continents to drift over the earth's surface.

Wetland: A place where there is open water, or waterlogged soil, for all or part of the year. Lakes, rivers, marshes and bogs are all wetlands.

Other Books to Read

There are many good books about our earth for young readers. Here are just a few of them:

Series:
Eyewitness Books.
London: Dorling Kindersley.
There are many books in the Eyewitness series about the earth and its life. The Eyewitness Science series includes: *Earth*; *Ecology*; *Evolution*; and *Life*.

Great Creatures of the World.
Sydney: Weldon Owen.
This is a series of very good books adapted for children from another series for adults. Titles include: *Sharks*; *Alligators and Crocodiles*; *Dolphins and Porpoises*; *Whales*; and *Elephants*.

Natural History Series.
Toronto: Key Porter Books.
San Francisco: Sierra Club Books.
Titles so far in this series include: *Eagles*; *Wolves*; *Bears*; *Seals*; *Snakes*; and *Apes*.

Individual Titles:
Bailey, Joseph H.
Giants from the Past: the Age of Mammals.
Washington, DC: National Geographic Society, 1983.
The story of extinct horses, rhinos, elephants, big cats and other mammals, and how we have learned about them.

Cullis-Suzuki, Severn.
Tell the World.
Toronto: Doubleday, 1993.
Severn Cullis-Suzuki tells about giving her speech, at the age of eleven, at the 1992 Rio Conference on Environment and Development in Rio de Janeiro, Brazil.

Curtis, Neil and Michael Appleby.
Planet Earth.
New York: Kingfisher Books, 1993.
One of the Visual Factfinder series, with a lot of information about the earth and how it works.

Deigler, Teri.
The Canadian Junior Green Guide.
Toronto: McClelland & Stewart, 1990.
This book explains what we are doing to our environment and what you can do to help save it.

Dixon, Dougal.
Dougal Dixon's Dinosaurs.
Honesdale, Pennsylvania: Boyds Mills Press, 1993.
This book about dinosaurs and their lives has been recommended by The Dinosaur Society.

Earthworks Group.
50 Simple Things Kids Can Do to Save the Earth.
Kansas City: Andrews and McMeel, 1990.
Here are things you can do every day to make this a better planet for all of us to live in, animals and plants included!

Farndon, John.
How the Earth Works.
London: Dorling Kindersley, 1991.
One hundred experiments you can do to learn more about the earth.

Horner, John R. and Don Lessem.
Digging Up Tyrannosaurus Rex.
New York: Crown Publishers, 1992.
A famous dinosaur expert tells how he and his team dug up the most complete *T. rex* specimen ever found.

Peters, David.
Giants of Land, Sea and Air, Past and Present.
New York: Alfred A. Knopf, 1986.
A panorama of the largest animals on earth, living and extinct, beautifully painted to scale next to running (or swimming) human beings. The blue whale stretches across seven fold-out pages!

VanCleave, Janice.
Earth Science for Every Kid: 101 Easy Experiments that Really Work.
New York: John Wiley and Sons, 1991.
Experiments on everything from mountain-building to tornado-making — in miniature, of course!

Wyatt, Valerie.
Weather Watch.
Toronto: Kids Can Press, 1990.
A book of activities and experiments to help you understand the weather.

Index

Numbers in *italics* refer to photographs or illustrations.